Sven Müller

Dynamic School Network Planning in Urban Areas

Stadt- und Raumplanung

Band 5

LIT

Sven Müller

Dynamic School Network Planning in Urban Areas

A multi-period, cost-minimizing location planning approach with respect to flexible substitution patterns of facilities

LIT

Bibliographic information published by the Deutsche Nationalbibliothek
The Deutsche Nationalbibliothek lists this publication in the Deutsche Nationalbibliografie; detailed bibliographic data are available in the Internet at http://dnb.d-nb.de.

ISBN 978-3-8258-1805-0
Zugl.: Dresden, Technische Universität, Diss., 2008

A catalogue record for this book is available from the British Library

Verlagskontakt:
Fresnostr. 2 D-48159 Münster
Tel. +49 (0) 2 51/620 32 - 22 Fax +49 (0) 2 51/922 60 99
e-Mail: lit@lit-verlag.de http://www.lit-verlag.de

Auslieferung:
Deutschland/Schweiz: LIT Verlag Fresnostr. 2, D-48159 Münster
Tel. +49 (0) 2 51/620 32 - 22, Fax +49 (0) 2 51/922 60 99, e-Mail: vertrieb@lit-verlag.de

Österreich: Medienlogistik Pichler-ÖBZ GmbH & Co KG
IZ-NÖ, Süd, Straße 1, Objekt 34, A-2355 Wiener Neudorf
Tel. +43 (0) 2236/63 535-290, +43 (0) 2236/63 535 - 243, mlo@medien-logistik.at

Distributed in the UK by: Global Book Marketing, 99B Wallis Rd, London, E9 5LN
Phone: +44 (0) 20 8533 5800 – Fax: +44 (0) 1600 775 663
http://www.centralbooks.co.uk/html

Distributed in North America by:

Transaction Publishers
New Brunswick (U.S.A.) and London (U.K.)

Transaction Publishers
Rutgers University
35 Berrue Circle
Piscataway, NJ 08854

Phone: +1 (732) 445 - 2280
Fax: + 1 (732) 445 - 3138
for orders (U. S. only):
toll free (888) 999 - 6778
e-mail:
orders@transactionspub.com

to my parents

Acknowledgements

There are many people to thank for their support and encouragement, without whom this thesis would not have been possible.
Firstly I thank my supervisor, Prof. Dr. Knut Haase. His continual support and encouragement have kept me going over the last four years. Anytime, he provided me helpful remarks and guidance for realizing this work. I would like to thank Prof. Dr. Georg Hirte for his helpful comments on the theory and applications of discrete choice methods and Prof. Dr. Gerhard Bahrenberg for supervising this thesis. Thanks also to the team members of the Institute for Transport and Economics at the Technische Universität Dresden, particulary Andrea Förster, Michael Klier, Andras Kéri, Stephan Brunow and Mirko Hope. I would like to thank all students for their supportive work, especially Stefan Tscharaktschiew and Sascha Kless. The helpful discussions at the Institute of Transport Studies and the School of Geography at the University of Leeds are gratefully acknowledged, too. For financial support I thank the Dresdner Verkehrsbetriebe AG (DVB AG), Deutsche Forschungsgemeinschaft (DFG) and the Gesellschaft von Freunden und Förderern der Technischen Universität Dresden e.V.
To my family I would like to thank for their great support and love. Finally to my wife, Eva, thank you for your love and support and for your unshakeable belief that no matter what new hurdle appears, we would overcome.

Abstract

Within a spatial competitive framework decision makers of services frequently face the question what is the optimal point in time to open or close a facility location under probabilistic and dynamic demand? Although this question has a long term tradition in human geography (Weber 1909, Lloyd and Dicken 1990, Drezner and Hamacher 2001) it is still difficult to answer since demand is usually not monotonic and often there are substitution patterns between different facility locations.
This thesis presents a dynamic cost-minimizing location planning approach which considers the availability of service locations to customers and the allocation of demand according to utility-maximization and substitution patterns. We apply this approach to find an optimal school net of private and public colleges over a long planning horizon in an urbanized area. Assuming a competitive market for educational services, decision makers are able to determine

(a) the appropriate point in time to enter or exit the geographic market,

(b) the expected market share as well as the transport costs, the fix and variable location costs and

(c) the optimal facility location(s) for market entry or exit respectively.

Following the outlined strategy, decision makers can successfully enhance their competitive position through dynamic location decisions.

Contents

List of Figures

List of Tables

Nomenclature

The abbreviations appear in alphabetical order while the symbols are sorted by chapter. If a symbol is used in more than one chapter it appears in the nomenclature in the chapter where it is used first.

Abbrevations

AMD	Advanced Micro Devices, Inc.
CED	Census Enumeration District
Cov	Covariance
Cplex	Optimization software package
CPU	Central Processing Unit
DMCFLP	Dynamic Modular Capacitated Facility Location Problem
DSDPP	Dynamic School District Planning Problem
DSNPP	Dynamic School Network Planning Problem
ESML	Exogenous Sampling Maximum Likelihood Estimator
GAMS	General Algebraic Modelling System
GIS	Geographic Information System
GP	Goal Programming
IIA	Independence from irrelevant alternatives
IID	Independent and identically distributed
LP	Linear Program
MINLP	Mixed Integer Non Linear Program
MIP	Mixed Integer Program

MMNL	Mixed Multinomial Logit Model
MNL	Multinomial Logit Model
MSLE	Maximum Simulated Likelihood Estimator
NFP	Network Flow Problem
OLS	Ordinary Least Squares
QCP	Quadratic Constrained Program
SE	Standard Error
SBS	School Boundary Stability Problem Over Time
SDPP	School District Planning Problem
SLL	Simulated Log-Likelihood
SNPP	School Network Planning Problem
Trans	Transportation Problem
Wald	Wald-statistic
WESML	Weighted Exogenous Sampling Maximum Likelihood Estimator

Chapter 2

k	Student
K	Set of all students within the sample
s	School
S	Set of all (available) schools
r	Superscript indicating the current draw
R	Number of draws
θ	Parameters characterizing density function $f(\cdot)$
β	Random realizations from density function $f(\cdot)$
γ	Vector of coefficients of the systematic component of utility
μ	Random coefficient vector with zero mean
q_{ks}	Vector of observed variables
a_{ks}	Vector of observed variables
b_{ks}	Vector of observed variables (subvector of a_{ks}): error components
E	Expectation value
W	Covariance matrix

V_s	Size of school s (ordinal measure)
d_{ks}	distance travelled to school s by student k
C_s	Centrality of school s (distance to city center)
ε_{ks}	Stochastic utility component which is IID
ξ_{ks}	Stochastic utility component
o_{ks}	Variable, equals one if student k chooses school s (equals zero otherwise)
$f(\cdot)$	Density function characterized by a vector of underlying moment parameters
U_{ks}	Utility function for student k and school s
L_{ks}	Conditional (or logit) probability that student k chooses school s
π_{ks}	Unconditional probability that student k chooses school s
$\breve{\pi}_{ks}$	Unbiased estimator of π_{ks}

Chapter 3

m	Transport mode for the journey to school
M	Set of different transport modes
v_{km}	Deterministic utility component
ϵ_{km}	Random utility component (IID)
F_{km}	Utility of a transport mode m to a student k
ρ_{km}	Probability that transport mode m is chosen by student k

Chapter 4

t	Period
T	Set of periods
i	Demand point
I	Set of demand points
j	Scenario
J	Set of scenarios
S_j	Set of available schools s embedded in scenario j
J_t	Set of all feasible scenarios j for period t

J_{st}	Subset of all scenarios $j \in J_t$ which contain school s
n_i	Number of students located in demand point i
p_{is}	Empirical expected share of students located in demand point i attending an available school s
α	Weighting parameter for under-capacity
λ	Weighting parameter for deviation from empirical expected student share
G_s	Capacity of school s
H_s	Accepted amount of students exceeding capacity of school s
c_{tj}	Costs of scenario $j \in J_t$ for period t
c^+_{st}	Costs of opening school s in period t
c^-_{st}	Costs of closing school s in period t
w_s	Variable, denoting the number of students exceeding the capacity of school s
x_{is}	Variable, denoting the share of students located in demand point i to enroll on school s
z^+_{is}	Variable, denoting the positive deviation between x_{is} and p_{is}
z^-_{is}	Variable, denoting the negative deviation between x_{is} and p_{is}
x^+_{st}	Variable, equals one if school s is opened in period t (equals zero otherwise)
x^-_{st}	Variable, equals one if school s is closed in period t (equals zero otherwise)
y_{tj}	Variable, equals one if scenario $j \in J_t$ is chosen in period t (equals zero otherwise)

Nomenclature

1 Introduction

1.1 Purpose

More and more regions in Germany are confronted with declining school enrollment caused by a decreasing fertility rate and negative net migration. Within the aftermath of the german reunification in 1990 a remarkable east-west migration has taken place. Particularly, younger, high-educated families migrated from the former German Democratic Republic to the western parts of Germany. At the same time the fertility rate declined. This in turn implies the necessity to close some school locations (Peter 2005, Richter 2005). However, in some prospering regions - particularly urban areas - population and thus enrollment increases after periods of decline. While making decisions as to which schools should be closed or opened at certain points in time, authorities should be aware of spatial-temporal and socio-economic consequences (Peter 2004, Klameth 2005). Different generations of students have to choose schools from different sets of alternatives.
German students are free to choose a secondary school at which to enroll. That means enrollment is not determined by location of the students, because school districts do not exist. Within the framework of "free school choice" it is valuable to know which factors (combination of profiles for instance) increases the choice probability of a certain school and whether there is heterogeneity in the sensitivity of individuals to the influencing factors. Particularly in a spatial choice situation one has to consider

flexible substitution patterns between different alternatives (Haynes et al. 1988), notably if the choice set differs over time. With declining enrollment the competition between schools will increase. Easily accessible and attractive schools will gain more students. This in turn leads to a reduced risk of being closed by school authorities.
With less schools available students may face a longer journey-to-school (Heitkamp 2005). Since distance greatly influences the travel-to-school mode choice, students switch from modes appropriate for short distances like cycling to modes appropriate for longer distances like public transport (modal-shift). Latest studies on travel-to-school mode choice stress the preservation of neighborhood schools and thus the preponderance of activity-related commuting modes like walking or biking due to short distances (Ewing et al. 2005, de Boer 2005). Not least of all, these modes are beneficial for students' health (McDonald 2005, McMillan 2003) and they are related to considerably lower costs in contrast to public transport or car. The closure of schools leads to cost savings for authorities in infrastructural and personnel. But there could be an increase in transport and social costs, which might yield increased total costs.
Concerning the school network planning there are numerous possible solutions for a given number of schools, demand points (blocks or districts for example) and periods. It will be nearly impossible to manually find an optimal school network related to a certain objective. As stated by Schlottmann (2005):

> *"Kultusminister Steffen Flath sorgt sich um die dauerhafte Stabilität des Schulnetzes. [..] 'Diese Entwicklung lässt befürchten, dass wir mit der Schulnetzplanung nicht mehr fertig werden' kritisierte Kultusminister Steffen Flath [..]"*

it is very difficult to adjust a school network manually. In order to handle this complex task, mathematical programming approaches are required. Although very sophisticated to some extent, more and more models are applied by authorities (November et al. 1996, Taylor et al. 1999, Caro et al. 2004, Tsai and Miller 2005, Teixeira et al. 2007). The literature mostly ignores free school choice within the context of school network planning (Lemberg and Church 2000). Nevertheless, it is evident that in recent years there is an emerging need for free school choice in countries other than Germany (Hoxby 2003, Burgess and Briggs 2006).

1.2 Concept

To analyze the issues mentioned in Section 1.1 we consider the concept shown by Figure 1.1. For this analysis a comprehensive survey has been carried out in order to have a unique data set at hand concerning the choice behavior (Chapter 2). Since we want to apply the approach on a geographical small scale, we have to disaggregate the data as far as possible. Therefore a method of data disaggregation and geocoding has been developed. According to the four-stage model (Bates 2000) we discuss the school choice first, followed by the travel-to-school mode choice. We use a mixed multinomial logit model (MMNL) in order to identify influencing factors of school choice and to regard realistic substitution patterns within the context of forecasting school choice probabilities with changing choice sets over time (Chapter 3). In Chapter 4 the travel-to-school mode choice is determined by a multinomial logit model (MNL). In order to incorporate the students' preferences of schools within the school network planning approach we have to consider two steps: First, for each period, we generate a set of scenarios indicating which school is open and which is not. For each scenario we allocate the students to available schools according to capacity and students' utility. The result of step one are the scenario costs (including transport costs determined by the MNL) for each feasible combination of scenarios and periods. Second, we select for each period one scenario that minimizes total costs (Chapter 5). Problems with large planning horizons and a geographical small scale are solvable. We apply this approach to the school network of the City of Dresden, Saxony (Chapter 6). A conclusion can be found in Chapter 7.

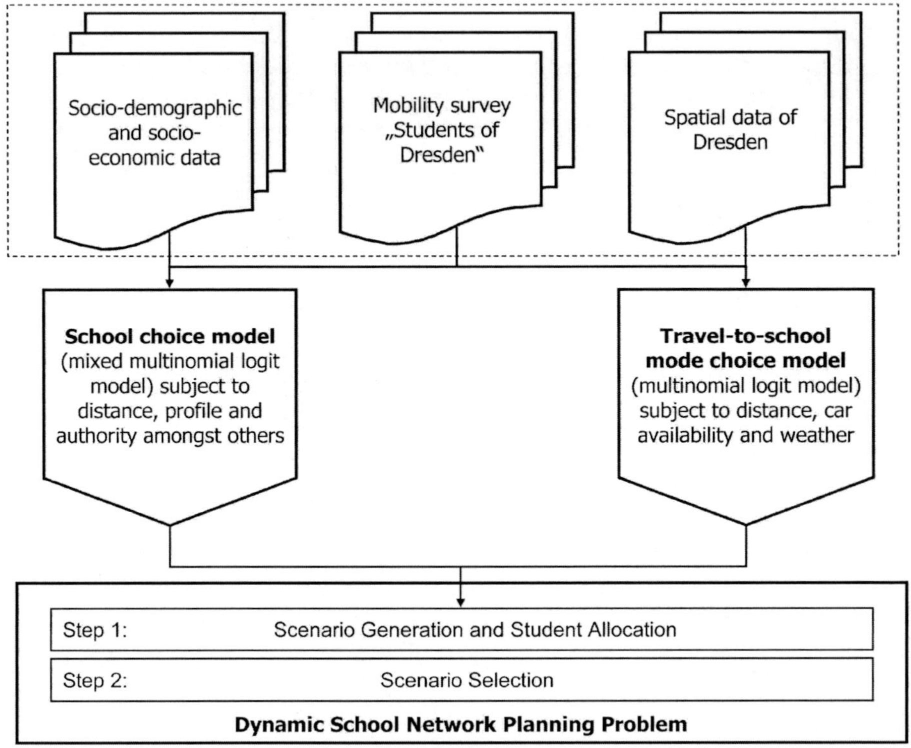

Figure 1.1: Structure of the Dynamic School Network Planning Approach

1.3 Additional Remarks

This thesis is based on three papers written by the author between 2005 and 2007.[1] The first paper *Travel-to-school mode choice modelling and patterns of school choice in urban areas* (Müller et al. 2007b) contains

[1]see http://tu-dresden.de/Members/sven.mueller/Research?set_language=en&cl=en

the disaggregation of the data, the travel-to-school mode choice modelling and a first insight into the topic of school choice. The second paper *Simulating school choice probabilities using a mixed multinomial logit model* (Müller and Haase 2007) deals with the MMNL of school choice behavior while the third paper *A dynamic school network planning approach with free school choice* (Müller et al. 2007a) contains the dynamic school network planning problem. The structure of this thesis is based on the content of the three papers: The introduction is more general while each chapter provides a specific introduction to the particular topic. Each model is based on the same data, but depending on the context, there are particular requirements on the data. So, Chapter 2 contains more general information about the data. A detailed description of the relevant data is given in each following chapter. The example of Chapter 4 (Section 4.2) does not consider the school choice model based on the MMNL, because the example is based on the paper *Travel-to-school mode choice modelling and patterns of school choice in urban areas.* The specified MMNL was not available at the time of the papers preparation.

2 Mobility Survey "Students of Dresden"

In this chapter we depict the design of the empirical survey, what data are available and how these data are disaggregated using a Geographic Information System (GIS). We do not discuss the data in detail - within the meaning of descriptive statistics. This is done in Section 3.2.2 and Section 4.1.1. In Section 2.1 we generally describe the data used and present a method of data disaggregation, followed by a short outlook on possible future research (Section 2.2).

2.1 Data and Disaggregation

2.1.1 General Data

This study is focused on secondary schools, particularly colleges (german: "Gymnasium"). College students are usually aged between 10 and 19 years (see Figure 2.1). We use population data which cover the age groups 10-19 years on block level for the years 1995 to 2020 provided by the local statistics authority. In Dresden around 45 per cent of all secondary school students are college students (City Council of Dresden (=Landeshauptstadt Dresden) 2003). The possibility to enroll at a college depends on the elementary school report (overall average grade). The number of 5th grade college students per block and year are based on these data. Our data set includes administrative areas, the school locations, the street network, the bus and tram stops, and the routes of the public transportation system of Dresden. As administrative areas we consider districts and blocks.[1] A block is bordered by streets (see Figure 2.2). Note, each district consists of a unique set of blocks. Using a shortest path approach we determine the street network distances between all blocks within Dresden. This includes the street network distances from each block to 26 college locations. These distances have to be interpreted as walking distances, since we do not have information about accessibility for cars, like one-way streets for instance. Here we just consider the journey-to-school and hence cars or motorcycles do not play an important role (see also Section 4.1).

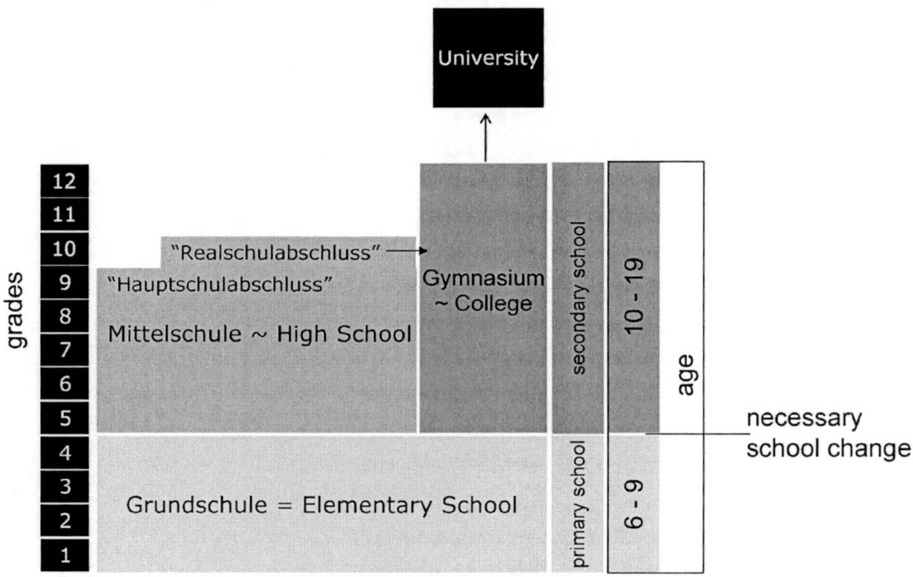

Figure 2.1: Main Features of the Educational System of Saxony, Germany

[1]There are 64 districts and more than 6400 blocks in Dresden.

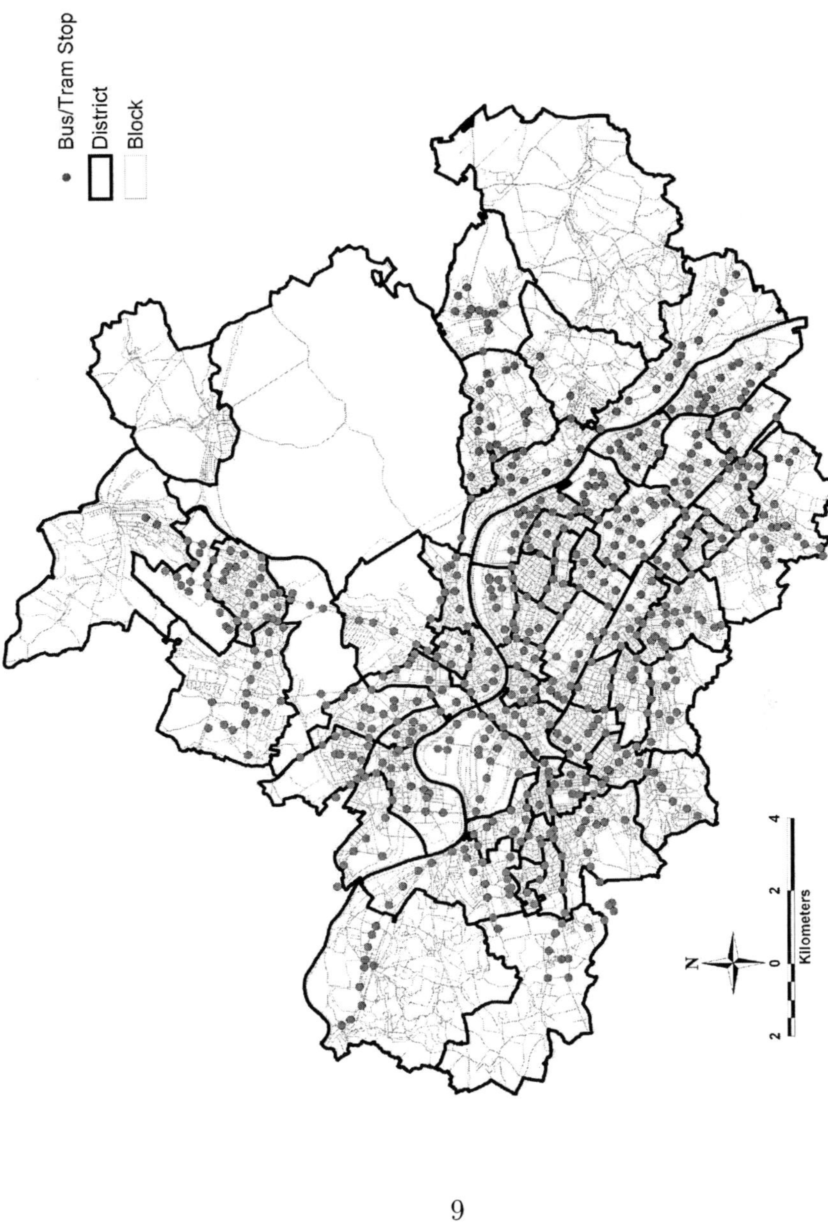

Figure 2.2: The City of Dresden - Administrative Areas and Public Transport Access. Blocks are Surrounded by Streets

2.1.2 Survey Data

In 2004, a survey covering nearly 8000 of 14 000 college students at 15 of 23 colleges in Dresden was carried out by the Chair of Business Administration, Transport and Logistics (University of Dresden). The whole survey lasted from January to November including a pre-test. We used a short form questionnaire which is similar to the one used by the German Federal Ministry of Transport for analyzing mobility behavior (Federal Ministry of Transport, Building and Urban Affairs (=Bundesministerium für Verkehr, Bau und Stadtentwicklung (BMVBS)) 2002).[1] Information on each student are the home district, the attended school, age, sex, car availability and whether the student owns a driver's license as well as commuting mode and total travel-time. Unfortunately, we only have information as to which school a given student attends, we do not know which school the student prefers. We assume that a student who is not allowed to attend the school with the highest utility - due to capacity reasons for example - is willing to attend the school with second highest utility and so on. So, for every student our sample provides information about the chosen school, which is the "allowed" school to enroll on with highest utility.[2] The total travel-time is related to the most preferred commuting mode in the summer term. Students were asked to state their preferred commuting mode on the journey-to-school both in winter and summer term. We suppose that fair weather is synonymic to the summer term and bad weather to the winter term respectively (see Figure 2.3). Furthermore, the students were asked how often they use a certain mode while commuting to school within a representative week. Again, this information is available for the summer and winter term. In case of the usage of public transportation, there is information about bus and tram routes and stops (origin, destination and interchange). Moreover, the students were asked to state their waiting times (departure station,

[1] The questionnaire is available on request.

[2] Note that it is supposable that the decision which school to choose is mainly influenced by parents.

interchange) and access as well as egress times, which are the walking time from home to the departure station and the walking time from the destination station to school. The questionnaire ends with questions, amongst others, on the ticket used and the satisfaction with the level of service.

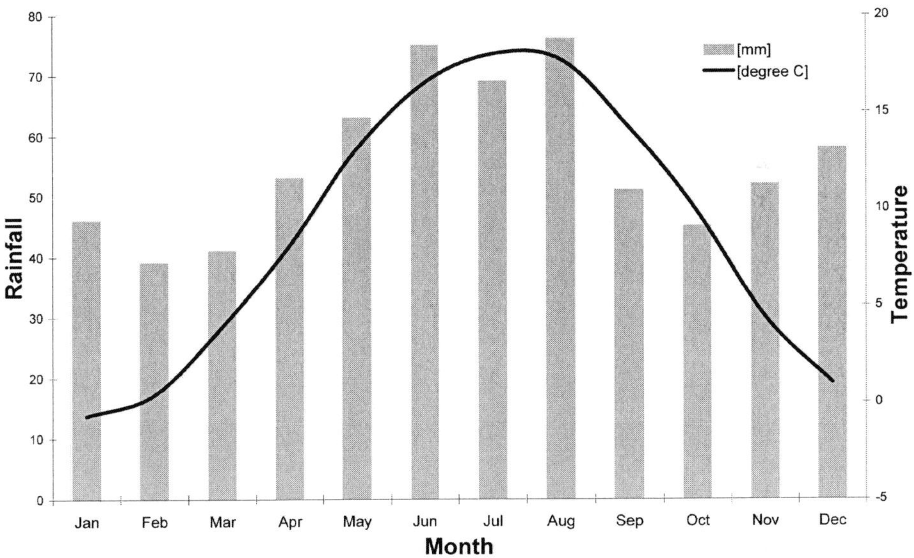

Figure 2.3: Climate Diagram of Dresden, Saxony

2.1.3 Disaggregation

Due to administrative restrictions which prohibit inquiring about detailed student addresses, we were forced to design a method for small scale (blocks) geocoding of the survey data using a GIS. The data were collected on the scale of districts. Since distance is an important variable for school choice as well as for the discrimination between most of the transport modes, data at the most disaggregated scale is needed in order to obtain a good approximation of exact distances for each student. Several authors stress the use of disaggregated data for distance related analysis (Goodchild 1979, Bach 1981, Fotheringham et al. 1995, Longley et al. 2001). There are sparse methods covering data disaggregation for transport surveys, but some work has been done in other fields of research (Gimona et al. 2000, Spiekermann and Wegener 2000, Van der Horst 2002, Greaves 2004, Oosterhaven 2005). Due to the extent of the survey sample and time limitations a subset of 4700 students were geocoded.

Most of the students use public transportation on their journey-to-school (50-60 per cent, see Section 4.1). Thus, the used departure bus or tram stop and the walking time required for the commute is known. Now, let us assume a student is located in district A (see Figure 2.4). Taking into account an average walking speed of 4 km/h, one can determine a student specific isochrone around the stated departure bus or tram stop. So, just a few blocks possibly contain the home of the student. The number of possible blocks could be reduced by considering the bus or tram route chosen by the student. This is based on the assumption that most of the students use the bus stop of the chosen line which is closest to their home. However, sometimes more than one possible block has to be taken into account for allocating the specific student although using all information available. Students with comparable properties (travel time, home district for example) are allocated to the considered blocks relative to the population of the specific age-group.

Regarding students who never commute to school by public transportation this detailed information is not available. In this case the following procedure has to be used: Imagine another student living in district A and the attended school is located in district B (see Figure 2.5). Again, the information of the commuting mode is available from our survey data, as well as the total travel-time. We assume a transport mode specific av-

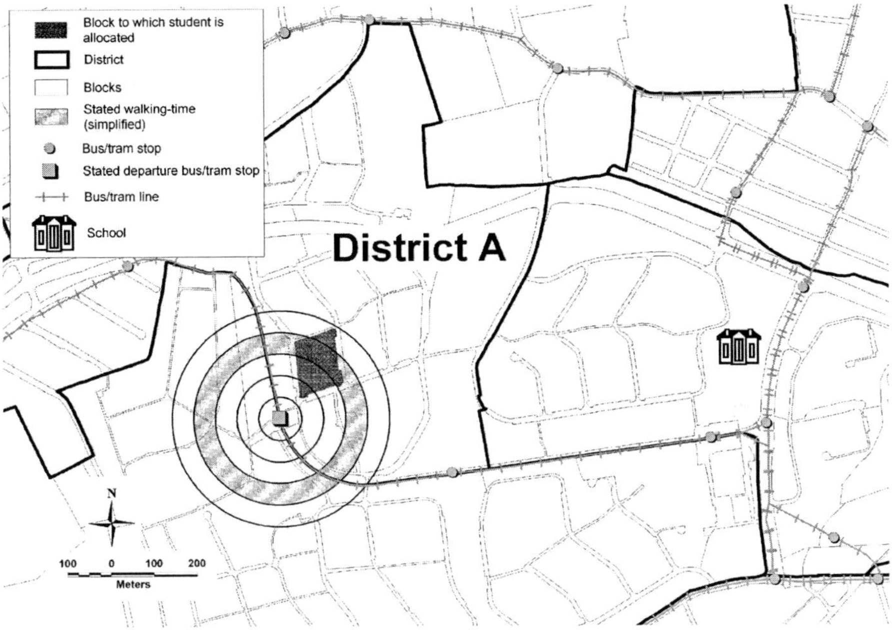

Figure 2.4: Allocation of Students Using Public Transportation

erage speed for walking of 4 km/h and for cycling of 12 km/h (Federal Environment Agency Germany (=Bundesumweltamt) 2007). The speed limit for cars and motorcycles is usually 50 km/h. Due to traffic lights and congestion we suppose an average speed of 30 km/h for cars and motorcycles in Dresden (German Aerospace Center (=Deutsches Zentrum für Luft- und Raumfahrt (DLR)) 2007). We expect these average speeds to be sufficient for the geocoding process. Using the average speed and the stated travel-times, we are able to determine a student specific isochrone around the attended school. For the modes biking and in particular car/motorcycling these isochrones are larger than those around bus stops (see above). This means that, there is more uncertainty about the correctness of the allocation of students to blocks. However, there is a very small percentage (6-10 per cent) of students who commute to school by car or motorcycle (see Section 4.1), and we expect that possible errors will be limited due to the extent of the sample.

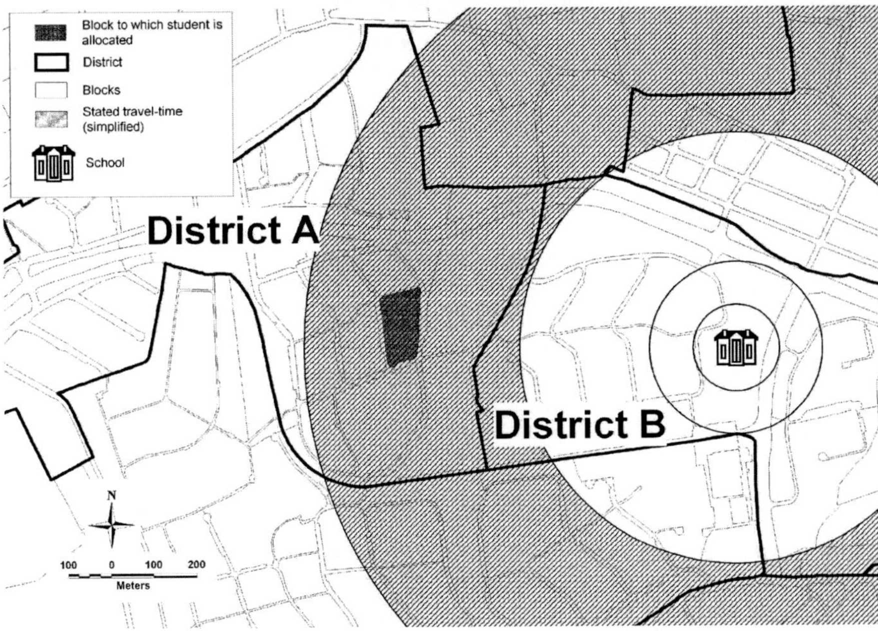

Figure 2.5: Allocation of Students **Not** Using Public Transportation

2.2 Outlook

We discuss a method utilizing GIS to disaggregate travel survey data. Particularly for the analysis of the journey-to-school this could be a useful procedure to gain better and even more realistic modelling results. Mostly, students choose public transportation and thus detailed spatial information are available. Future research should incorporate the outlined procedure into a standard GIS.

3 School Choice Modelling

In this chapter we utilize a MMNL to describe the process of school choice and simulate the choice probabilities related to the opening and closing of schools in the City of Dresden, Saxony. We expect that commuting distance and the school specific characteristics (profiles for example) are important and sufficient factors determining the choice of a certain school location in Saxony, Germany (Speiser 1993, Mahr-George 1999, Schneider 2004, Müller 2006, Müller et al. 2007b). We differentiate between schools with a common profile and schools with a unique profile. A school with a common profile provides a service (mathematics/science for example) which is comparable to a lot of other schools. Usually, schools offer a combination of two common profiles. In contrast, a unique profile is offered by one school only. For example: If a school offers far more languages than most of the other schools we name this unique profile *core languages*. Moreover, we name a school offering a unique profile a *magnet school* since it is the only one in a particular region offering such an educational service. So students are expected to be attracted by a magnet school over a broad geography. Note that tuition fees only occur at some private schools in Dresden. They are at a low level usually. So, we do not explicitly pay respect to tuition fees in this study. Also, the confessional affiliation is neglected here as an explanatory variable. There are no confessional constraints concerning enrollment and again just a few private schools have a confessional affiliation. However, the number of private schools is increasing recently. In future, tuition fees and confessional affiliation are expected to have a much greater impact

on school choice (Barthels 2007b).
The chapter is structured as follows: A general analysis of the patterns of school choice is given in Section 3.1. Section 3.2 covers an overview of the literature and the theoretical model determining the school choice probabilities. In Section 3.2.2 we describe the data used followed by the estimation of a series of models (Section 3.2.4). In Section 3.3 we simulate the choice probabilities for different scenarios, which cover school openings and closures. Some final remarks can be found in Section 3.4.

3.1 Patterns of School Choice

Several surveys yield proximity and the authority responsible (private or public school) as two very important factors of school choice. Other factors are the reputation of schools and tuition fees for example (Speiser 1993, Mahr-George 1999, Hoxby 2003, Schneider 2004, Hastings et al. 2005). We expect that the offered profiles could have influence on school choice as well. In this study we focus on distance, the school's profiles, and the authority responsible to determine the school choice, since most of the other influencing factors stated in the literature cannot be applied here due to the lack of data or the relative unimportance of the variable. For instance, the willingness-to-pay for schools measured as the ratio of household income and tuition fees is an important economic factor influencing school choice which is considered in some of the studies mentioned above. We consider the german educational system and at the moment tuition fees are an exceptional case. Moreover there is no detailed information on the average household income available. For an overview of school locations and profiles offered see Figure 3.1.

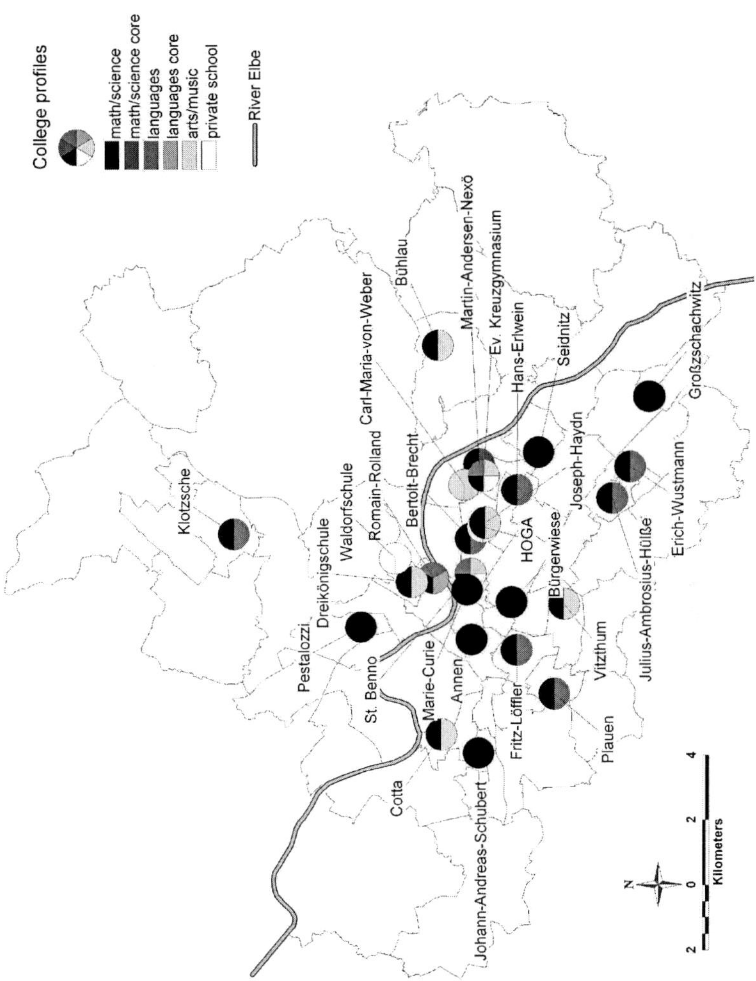

Figure 3.1: Colleges in Dresden and Offered Profiles

3.1.1 School Catchment Area and Proximity

First, we have to determine the surrounding catchment area of each school. Therefore, the nearest school location has to be verified for each block. Because students will not always realize this strictly drawn border, we add two zones with reduced distances (zone 2: -1000m and zone 3: -2000m). Consequently, the distances of blocks within zone 2 and 3 are minimal to the specific school location (see Figure 3.2). Table 3.1 shows the percentage of students within the corresponding zones for all schools of in sample. For example, 84.8 per cent of all students attending Klotzsche college are located in zone 1. The surrounding catchment area of each school consists of three zones as defined above. We believe that within this area students recognize the specific school as the closest one. Two main patterns are evident:

- Students attending schools with a common profile mostly are located in the surrounding catchment area. Thus, one can assume that proximity is an important factor for school choice.
- For schools with a unique profile and for private schools this does not hold true. It appears to be that a unique profile, or a private school, conflicts with the importance of proximity.

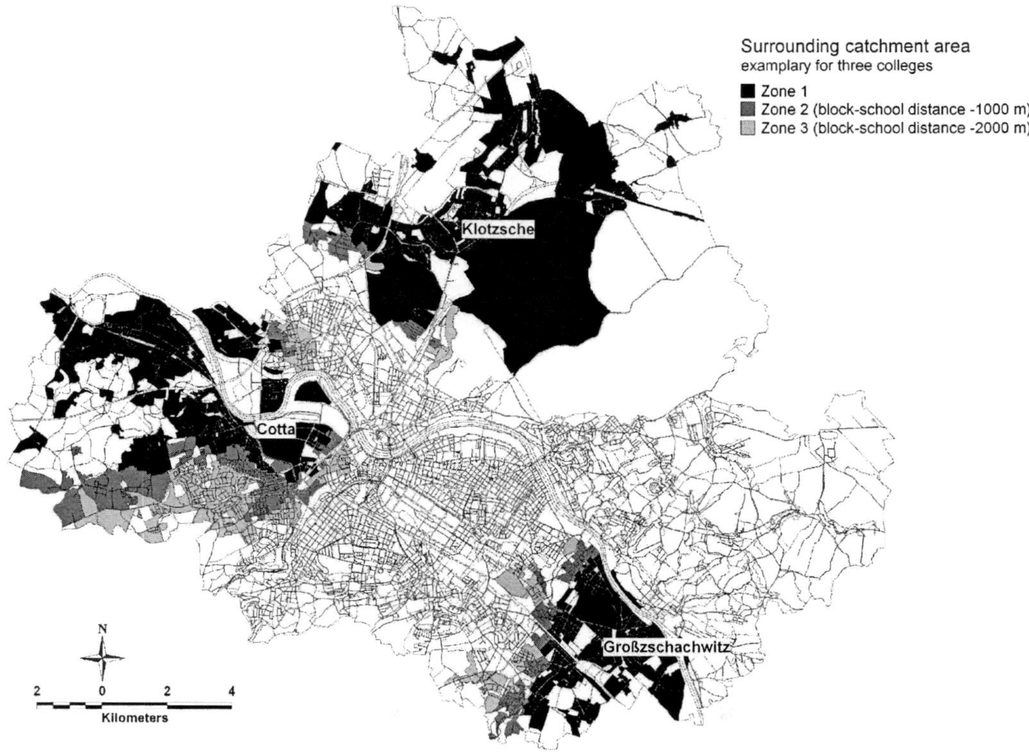

Figure 3.2: Surrounding Catchment Areas: Populated Blocks to Which One of the Three Exemplary Colleges are the Closest One

College	Zone 1	Zone 2	Zone 3	Other	Sum Zone 1-3
Klotzsche	84.8	0.5	0	14.7	85.3
Großzschachwitz	68	12.9	1.7	17.4	82.6
Cotta	50	23.3	17.5	9.3	90.7
Julius-Ambrosius-Hülße	32.6	28.7	12.4	26.4	73.6
Fritz-Löffler	35.8	11.2	23	30	70
Plauen	46	9.4	8.9	35.7	64.3
Vitzthum	48.6	5.7	17.7	28	72
Romain-Rolland	9.4	18.8	3.6	68.2	31.8
Marie-Curie	8.3	14.8	11.6	65.3	34.7
Martin-Andersen-Nexö	15.1	26.1	17.6	41.2	58.8
Joseph-Haydn	7.3	17	32.1	43.6	56.4
St. Benno*	0.4	4.1	10.2	85.3	14.7

Only schools which are covered in the survey by at least 150 students are considered
Colleges written **bold** are magnet schools (unique profile)
* Private school

Table 3.1: Percentage of Students Located in School Specific Catchment Areas

Outliers in both groups - Marie-Curie College and Martin-Andersen-Nexö College for example - are due to the topology of the school network (see Figure 3.1). Schools located in an area with high density of school locations (spatial cluster) obtain smaller surrounding catchment areas and thus fewer students within. At the outskirts these catchment areas are larger and contain more students. For students located in blocks close to a spatial cluster it is not always obvious which school location is the closest one. Within a cluster there are many choice alternatives available within a reasonable proximity (Bertold-Brecht College for instance, see Figure 3.1). It is reasonable to assume that if the closest school does not match the preferences of students for a combination of profiles etc., a school within the spatial cluster does so. For students located in a spatial cluster proximity is less important than for those students located at the outskirts. At the same time, other properties, such as profile, are more important for the decision as to which school to enroll at. Therefore, the location of the school (central/peripheral) may have an additional influence on school choice.

3.1.2 Profile and School Choice

For a deeper investigation of the influence of profile and the authority responsible we consider Table 3.2. It shows the number of students who attend schools which are not the closest one. Over all colleges most of the students (80 per cent) choose a school with a different profile and/or a different authority responsible. Let us take Klotzsche College as an example: Klotzsche College is the closest one for 100 students of our sample, but these 100 students actually choose a different school (sum 1-5). 88 of them choose colleges with a different profile and/or a different authority responsible (sum 2-5). 20 of these 88 students choose colleges with an alternative profile (column 3). Over all colleges nearly 70 per cent of the students who choose a different school than the closest one, choose a school with a different profile. Therefore we assume that profile and the authority responsible are two factors which influence the choice of a certain school. The students who attend a school with the same properties as the closest one (20 per cent) may be attracted by factors not considered here (extracurricular program for example). Another possible reason may be inner-city migration.

3.2 School Choice Model

This section deals with the student's individual choice probability for each school, dependent on commuting distance and school specific characteristics. We assume that each student chooses the school with the highest utility. The multinomial logit model (MNL) has been well established within the utility maximization context of destination or location choice modelling (McFadden 1973, Ben-Akiva and Lerman 1985, Hansen 1987, Gabriel and Rosenthal 1989, Hunt and Teply 1993, Bhat 1997, Greene 2003, Becker et al. 2005). Recent studies on school choice - partly applying the MNL - are focused on the discrimination between the choice of private and public schools (Figlio and Stone 1997, Alderman et al. 2001) or the question whether students will take part in higher

students in zone 1 of	attending a college with 1	2	3	4	5	$\sum$ 2-5	$\sum$ 1-5	$\sum$ 2-5 / $\sum$1-5
Klotzsche	12	29	20	0	39	88	100	.88
Großzschachwitz	49	0	169	0	30	199	248	.80
Cotta	23	0	20	0	17	37	60	.62
JAH***	37	0	31	0	32	63	100	.63
Fritz-Löffler	0	8	81	0	11	100	100	1
Plauen	40	4	67	0	25	96	136	.71
Vitzthum	1	50	97	0	20	167	168	.99
Romain-Roll.	0	3	12	0	13	28	28	1
Marie-Curie	1	0	21	0	6	27	28	.96
MAN**	47	5	19	0	33	57	104	.55
Joseph-Haydn	23	0	20	0	16	36	59	.61
St. Benno*	0	0	0	24	0	24	24	1
Average (total)	19.4	8.3	46.4	2	20.2	76.8	96.3	.80
Average (%)	20	9	48	2	21	80	100	

1: identical profile and authority responsible
2: additional feature (additional language i.e.)
3: alternative profile
4: different authority responsible
5: alternative profile and different authority responsible

Colleges written **bold** are magnet schools
*: Private school
**: Martin-Andersen-Nexö
***: Julius-Ambrosius-Hülße

Table 3.2: School Choice of Students Choosing a Different School Than the Closest One

education. For example, whether they enroll at a college (Brinkmann and Leslie 1987, Kane and Spizman 1994, Schneider 2004). Lankford and Wyckoff (2000) study the relationship between racial composition of schools and school choice. Their results show that the racial composition of schools and neighborhoods are very important within the context of choosing between private and public schools. The choice between voucher and non-voucher schools in Chile has been studied by Contreras (2002). He reveals that school choice is gender based. Females more often choose voucher schools, while males more often choose private, non-voucher schools. The results of Nguyen and Taylor (2003) seem to evidence that a full choice set is superior to combinations of sub-sets of choice alternatives for the modelling of college choice. Montgomery (2002) uses the logit approach to model the process of university choice mostly using cost variables. Hoxby (2003) mainly discusses the economic benefits of free school choice such as increase in productivity and welfare but pays less attention to which variables determine the process of school choice. The most influencing factors for parental choice of a school seem to be proximity and academic quality (Goldring and Bauch 1995, Glazerman 1997, Kleitz et al. 1998, Peterson and Hassel 1998).

In Eastern German cities the share of ethnic minorities is very low (except Berlin) and so we assume that the racial composition does not have a strong influence on school choice here. There are no voucher schools in Germany and there are just a few private schools. So most of the studies previously mentioned focus on (or at least consider) attributes which seem to be of less interest concerning school choice in Germany and particularly Saxony. Moreover, it is very difficult to forecast some of the variables discussed in the literature, like ethnic shares for example. Based on the results of Section 3.1 we assume that school choice is related to profile, authority and proximity.

We model the school choice behavior in order to predict choice probabilities within scenarios of school closures and openings. Therefore we consider flexible substitution patterns between schools. Also, we explore taste variation within the population. Here the MNL is not appropriate because its restrictive assumptions concerning the stochastic utility component (Train 2003, p.46-56), for example it exhibits independence from irrelevant alternatives (IIA) and yields unrealistic substitution patterns. Concerning school choice modelling, Borgers et al. (1999) points out that a standard/conventional logit model is not appropriate to include substitution effects. In this case it appears to be that the more general MMNL is superior to MNL due to its flexibility concerning substitution patterns and individual heterogeneity (Brownstone and Train

1999, Dow and Endersby 2004, Gottlieb and Joseph 2006). Besides empirical experience with this kind of discrete choice model from other fields of research (Rouwendal and Meijer 2001, Hensher and Greene 2003, Bhat and Gossen 2004, Smith 2005) and particularly Hastings et al. (2005) use a MMNL to identify parental preferences and school competition. They analyze how demand disperses in a situation of free school choice for elementary and middle schools. Using data from a stated preference survey they discover proximity, ethnicity and academic quality as the most influencing factors. These variables show significant variation in individual heterogeneity.

3.2.1 Mixed Multinomial Logit Structure

The mixed logit probabilities are the integrals of the well-known logit probabilities L_{ks} over a density[1] of parameters (Bhat 2000). Generally, the model takes the structure:

$$\pi_{ks}(\theta) = \int L_{ks}(\beta) f(\beta|\theta) d(\beta) \tag{3.1}$$

with

$$L_{ks}(\beta) = \frac{\exp(\beta' q_{ks})}{\sum_{\tilde{s}} \exp(\beta' q_{k\tilde{s}})} \tag{3.2}$$

where π_{ks} is the probability that student k chooses school s. Here, q_{ks} is a vector of observed variables, while β represents parameters that are random realizations from a density function $f(\cdot)$ characterized by a vector of underlying moment parameters θ. β are normally distributed in our case. Here, the overall random term associated with the utility

[1]This density is also called the mixing distribution. The name mixed multinomial logit model is dispositional on this.

of each alternative is divided into two components. In order to capture individual heterogeneity and correlation across alternatives, the utility function U_{ks} for student k and school s is specified as:

$$U_{ks} = \gamma' a_{ks} + \xi_{ks} = \gamma' a_{ks} + \mu_n' b_{ks} + \varepsilon_{ks} \tag{3.3}$$

where a_{ks} and b_{ks} are vectors of observed variables related to school s and student k, γ is a vector of coefficients of the systematic component of utility, μ_n is a random vector with zero mean and ε_{ks} is independent and identically extreme value distributed (IID). The terms in b_{ks} are error components that, along with ε_{ks}, define the stochastic component of utility ξ_{ks}, which can be correlated over alternatives depending on the specification of b_{ks}. Here, b_{ks} is a subvector of a_{ks} primarily containing variables of the school attributes. The stochastic part of utility yields tastes not observed by the researcher are used by the student to evaluate schools (random taste variation), and since we do not observe these tastes completely, this portion of utility is correlated over schools (unobserved attributes). The covariance of two schools s and l is denoted by $\text{Cov}(\xi_{ks}, \xi_{kl}) = E(\mu_n' b_{ks} + \varepsilon_{ks})(\mu_n' b_{kl} + \varepsilon_{kl})$. We expect $E(\mu_n) = E(\varepsilon_{ks}) = E(\varepsilon_{kl}) = 0$. Let us denote the covariance of μ_n as W. So $E(\mu_n \mu_n') = W$. For a given n:

$$W = \begin{pmatrix} \sigma_{11} & \sigma_{12} & \sigma_{13} & \cdots \\ \sigma_{21} & \sigma_{22} & \sigma_{23} & \cdots \\ \sigma_{31} & \sigma_{32} & \sigma_{33} & \cdots \\ \cdots & \cdots & \cdots & \cdots \end{pmatrix}$$

with $\sigma_{11} = E(\mu_1 \mu_1)$ and $\sigma_{12} = E(\mu_1 \mu_2)$ and so on. If we consider the covariance of two different schools $\text{Cov}(\xi_{ks}, \xi_{kl})$, then

$$\begin{aligned}
& E(\mu_n' b_{ks} + \varepsilon_{ks})(\mu_n' b_{kl} + \varepsilon_{kl}) \\
= \; & E(\mu_n' b_{ks} \mu_n' b_{ls} + \mu_n' b_{ks} \varepsilon_{kl} + \mu_n' b_{kl} \varepsilon_{ks} + \varepsilon_{ks} \varepsilon_{kl}) \\
= \; & E(\mu_n' b_{ks} \mu_n' b_{kl}) \\
& \text{Since } b_{ks} \text{ and } b_{kl} \text{ are known rather than expected it follows} \\
= \; & b_{ks}' E(\mu_n \mu_n') b_{kl} \\
= \; & b_{ks}' W b_{kl}.
\end{aligned}$$

Hence, the MMNL handles the IIA problem automatically if the stochastic part of the utility function is specified in the way to capture the desired correlation. For instance, let us assume b_{ks} is a dummy variable which equals one if the considered school s is a private school. There would be a positive correlation if the other school l is a private school as well ($b_{kl} = 1$). Note, that the probabilities of equation 3.1 are approximated through simulation for any given value of θ (Train 2003). The process of estimation is outlined in Section 3.2.3.

3.2.2 Data for School Choice Modelling

Figure 3.3 displays the spatial distribution of all colleges and college students in Dresden for the years 2004, 2008 and 2012 (see Section 3.3). We do not consider the colleges Sportgymnasium and Abendgymnasium, because they are supra-regional colleges or rather special boarding schools. These schools attract students from the whole of Saxony. Consequently, they are not representative and would distort the modelling. The school size V_s, is the number of classes per grade (synonyms: year, class-level). Furthermore, we obtain information on the profiles offered by each college as well as the authority responsible. We assume that unobserved attributes of colleges are associated with the authority responsible. For example, it is likely to assume that private schools have a better academic quality than public schools because of a better teacher-student ratio. Private schools often have a confessional affiliation (Barthels 2007a). As mentioned in Section 3.2.1 the tastes of the students to these unobserved attributes are correlated over schools.

For each school our data set includes information on the size of a school s (V_s), profile, authority responsible (=1 private school, =0 public schools) and the distance d_{ks} travelled to school s by student k (see Table 3.3). We use distance instead of travel-time because we will consider future scenarios of the school network (see Section 3.3) and some of the school locations for future school network scenarios do not exist at the moment. Therefore, it is not possible to compute exact public transport travel-times for these school locations.[1] However, we assume a student evaluates distance with the travel-time of an appropriate commuting mode. So, in the choice situation distance serves as a proxy for travel-time implicitly.

Since we expect a nonlinear influence of distance on students school choice, we consider a log-distance measure besides the simple distance measure. It is expected that the location of a school in the meaning of centrality has influence on school choice as well. The centrality of a school C_s is measured in distance (km) from school s to the city center[2]. Schools located close to the city center are more easy to access by public transport than schools located at the outskirts. We assume that it is more attractive for a student located in an outskirt-district to attend a school located close to the city center (low value of C_s) than to attend a school located at a different outskirt-district (high value of C_s). We consider a dummy variable which indicates whether the closest school is a magnet school (=1) or not (=0). Regarding the average distance in Table 3.3 it is conjecturable that there will be a relatively high proportion of students choosing a magnet school even if located far away. Since the utility function is linear in parameters, enumeration maybe leads to

[1] One has to consider new lines and schedules for example.

[2] Here, the functional city center is located close to the topological centroid of the city area.

unrealisticly high utility for students located close to magnet schools - all other variables constant. Thus, the expected sign of this coefficient will be negative.
We have eliminated the colleges Großzschachwitz, Joseph-Haydn and Fritz-Löffler from the data set in order to avoid distortion. These colleges are already closed for enrollment in 2004 (year of the survey, see Chapter 2). For example, in 2004 there are students - especially at lower grades - attending Julius-Ambrosius-Hülße College which is next to Großzschachwitz just because they are not allowed to enroll at the closest college, Großzschachwitz (Müller et al. 2007b).

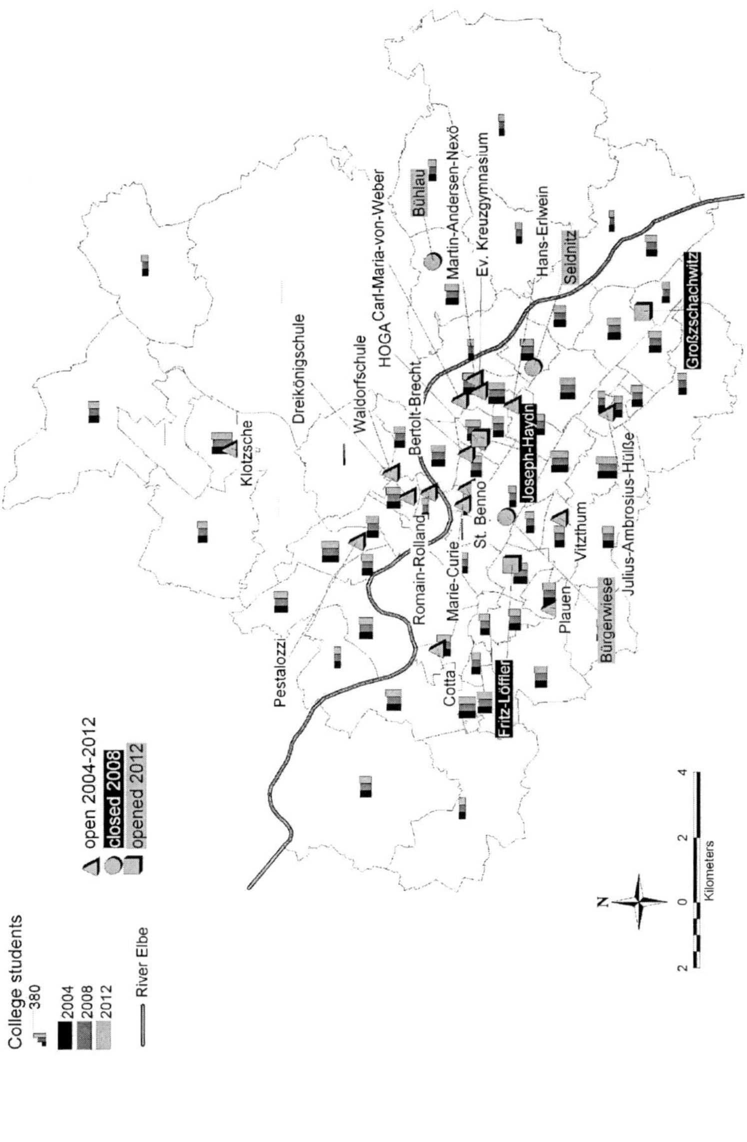

Figure 3.3: College Students and Colleges for the Years 2004, 2008 and 2012

College	private school	p1	p2	p3	p4	p5	# chosen altern.	factor WESML	distance [km] mean	SD	min.	max.	C_s [km]	V_s♠
Pestalozzi*	0	1	0	0	0	0							3.52	3
Großzschachwitz	0	1	0	0	0	0			2.69	2.58	.61	13.92	8.19	4
Marie-Curie	0	1	0	0	0	0	201	2.005	4.67	3.18	.10	16.12	0.39	4
Bürgerwiese**	0	1	0	0	0	0							1.14	3
Seidnitz**	0	1	0	0	0	0							4.94	3
Cotta	0	1	0	0	0	1	629	0.867	2.97	1.68	.10	9.49	4.22	5
Joseph-Haydn	0	1	0	0	0	1			5.03	3.11	.23	12.29	2.40	4
Dreikönigsschule	0	1	0	0	0	1	152	0.948	4.14	3.55	.07	12.09	1.98	3
Vitzthum	0	1	0	0	0	1	266	0.869	2.41	1.71	.07	12.46	2.75	4
Bühlau**	0	1	0	0	0	1							7.88	3
Bertold-Brecht*	0	1	0	1	0	0							1.90	4
Fritz-Löffler	0	1	0	1	0	0			2.41	2.26	.12	10.44	1.99	3
J.-A.-Hülße	0	1	0	1	0	0	620	0.853	2.50	1.96	.13	17.80	5.24	4
Klotzsche	0	1	0	1	0	0	343	1.116	2.24	1.96	.03	13.04	7.58	3
Plauen	0	1	0	1	0	0	441	1.099	2.25	1.87	.12	16.92	3.76	5
Hans-Erlwein*	0	1	0	1	0	0							3.63	5
Carl-M.-v.-Weber*	0	0	0	0	0	1							3.55	1
St. Benno	1	1	0	1	0	1	418	0.977	5.55	2.67	.15	14.68	0.83	3
Ev. Kreuzgymnasium*	1	1	0	1	0	1							3.83	3
MAN	0	1	1	0	0	0	261	0.961	3.82	2.57	.27	14.30	4.18	3
Romain-Rolland	0	0	0	1	1	0	373	0.867	4.65	3.09	.09	14.85	1.43	3
Waldorfschule*	1	0	0	0	0	0							2.73	2
HOGA*	1	0	0	0	0	0							2.27	1
average							370,4		3.38	2.64	.04	17.80	3.32	3.35

p1: math/science
p2: math/science core
p3: languages
p4: languages core
p5: arts/music

* colleges not within geocoded sample
** colleges opened in 2012
Colleges written **bold** are magnet schools
♠ number of classes per grade

Table 3.3: Data for School Choice Modelling

3.2.3 Estimation Process

The simulated school choice probabilities are computed as follows:

(1) Draw β from $f(\beta|\theta)$ and label it β^r with superscript $r = 1$ referring to the first draw.

(2) Calculate $L_{ks}(\beta^r)$ with this draw.

(3) Repeat steps 1 and 2 R times and average the results as $\breve{\pi}_{ks} = 1/R * \sum_r^R L_{ks}(\beta^r)$.

$\breve{\pi}_{ks}$ is the unconditional probability used to calculate the simulated likelihood function (SLL). $\breve{\pi}_{ks}$ is strictly positive and an unbiased estimator of π_{ks} by construction. Its variance decreases as R increases. As Bierlaire (2003) recommends we set $R = 1000$. We use Biogeme 1.4 for estimation with implemented Unix's pseudo random number generator. Other possible methods are Halton sequences and Modified Latin Hypercube Sampling (Hensher and Greene 2003). A deeper discussion is not given here. $\breve{\pi}_{ks}$ is smooth in θ and all variables and it sums to one over all alternatives. So the simulated probabilities are inserted into the log-likelihood function to give a simulated log likelihood:

$$SLL = \sum_{k=1}^{K} \sum_{s=1}^{S} o_{ks} \ln \breve{\pi}_{ks} \tag{3.4}$$

where $o_{ks} = 1$, if student k chooses school s and zero otherwise. The maximum simulated likelihood estimator (MSLE) are the values of β and θ that maximize SLL by iteration. The properties of this estimator are discussed in Train 2003, p.253-260. A trust-region algorithm (Conn et al. 2000) named BIO is used for optimization (Bierlaire 2005). The number of students attending a given school within our choice-based sample do not represent the actual numbers. So we have to use the weighted exogenous sampling maximum likelihood estimator (WESML) to obtain consistent estimators for the MMNL. Since we do not consider alternative specific constants we use the exogenous sampling maximum likelihood estimator (ESML) for the MNL. The author provides the data and the source code for the MNL and MMNL (exemplary) on request.

3.2.4 Results

We present and discuss the results of several model specifications. Although a Hausmann-McFadden test proves our suggestion that the IIA assumption holds not true for our sample, we estimate a MNL in order to show the advantages of the MMNL for this specific study. The choice of variables to enter utility were determined through exploration. It turns out that the dummy variable for profile mathematics/science is not significant because nearly every college offers this profile. The log of d_{ks} yields a bad model fit and we remove this variable. Experiments with different model specifications concerning the stochastic utility component show that profile music/arts has to be removed from the set of variables as well. This variable distorts modelling (high final gradient norm, non-convergence) using MMNL. On the basis of these pre-examinations, the variables distance, closest school is a magnet school (dummy), C_s, V_s, private school (dummy), profile 2, profile 3 and profile 4 (all dummies) seem to be most promising and have to be considered for estimation and choice probabilities simulation. All three MMNL models have to be interpreted within an error components structure motivation, which is formally equivalent to random components structure and differs only concerning the interpretation of the utility function (Train 2003, p.144). The three different model specifications (MMNL 1-3) are due to testing for heterogeneity in population and flexible substitution patterns (see table 3.4). In other words: Which variables enter the stochastic part of utility significantly? Surprisingly, the students tastes concerning the profiles do not vary. Pre-examinations show mostly insignificant standard deviation for the profiles. This suggests a distinct discrimination of tastes concerning school profiles by the dummy specification of profiles.

Variable		MNL Est.	MNL SE	MNL t-test	MMNL1 Est.	MMNL1 SE	MMNL1 t-test	MMNL2 Est.	MMNL2 SE	MMNL2 t-test	MMNL3 Est.	MMNL3 SE	MMNL3 t-test
Distance d_{ks}	mean	-.700	.014	-48.28	-.726	.014	-51.44	-.899	.034	-26.35	-.798	.019	-40.05
	Std. Dev.							.274	.045	6.09			
Centrality C_s	mean	-.043*	.031	-1.34	-.194	.021	-9.35	-.190	.028	-6.48	-.158	.026	-5.81
	Std. Dev.							.422	.040	10.40	.382	.039	9.65
Magnet school	mean	-1.150	.155	-7.42	-1.090	.157	-6.93	-1.676	.236	-7.11	-1.384	.190	-7.32
is closest	Std. Dev.												
Profile 2	mean	.496	.163	3.04	.415	.108	3.84	.459	.134	3.41	.435	.122	3.53
(math/science core)	Std. Dev.												
Profile 3	mean	.095*	.143	0.67	.222	.066	3.35	.200	.073	2.69	.180	.068	2.65
(languages)	Std. Dev.												
Profile 4	mean	1.190	.127	9.35	.504	.107	4.72	.656	.119	5.48	.634	.115	5.49
(languages core)	Std. Dev.												
Profile 5	mean	.144*	.115	1.26									
(arts/music)	Std. Dev.												
School size V_s	mean	.554	.049	11.31	.536	.047	11.24	.527	.056	9.42	.522	.055	9.57
	Std. Dev.												
Authority	mean	1.010	.221	4.59	-.736	.319	-2.31	-.762	.380	-2.00	-.660*	.367	-1.79
(=1, if private school)	Std. Dev.				2.280	.320	7.14	2.538	.402	6.29	2.358	.390	6.04
Log likelihood			-4641.13			-4846.02			-4800.74			-4813.57	
Likelihood ratio			7775.29			7365.50			7456.06			7430.41	
McFadden-R^2			.455			.431			.436			.434	
% correctly predicted			58.25			58.64			58.69			58.64	
All variables are significant on .95 level, except * (non-significant)													
The number of observations is 3704 for all models.													
Likelihood ratio test (χ^2)													
MNL			-			409.78			319.22			344.88	
MMNL1						-			90.56			64.91	
MMNL2									-			25.66	

Table 3.4: Estimation Results for Different School Choice Model Specifications

If we look at the four models we see that the coefficients of the spatial variables (d_{ks}, C_s and closest school is a magnet school) are larger for the MMNL models than for the standard logit model. The evaluation of the profiles differs between the logit model and the mixed logit models while there is no remarkable difference between the three mixed logit models. The coefficient for the school size is very similar over all four models. Authority responsible shows the most striking difference between MNL (positive sign) and MMNL (negative sign) specification.
The negative sign of distance indicates a decrease in utility with increasing commuting distance. MMNL2 discovers taste variation of students evaluating commuting distance. The negative sign for centrality C_s means that, other things being equal, colleges closer to the city center have a higher utility than those ones at the outskirts: it is most likely to assume that most individuals obtain highest utility from the closest school. Nevertheless, for some individuals the school characteristics (profiles, authority for instance) of the closest school might not be desirable and so they choose an alternative school. Mostly, this alternative school is not within walking or biking distance. The schools located near the city center have a good accessibility by public transport. Moreover, the high spatial density of school locations near the city center presents a lot of alternatives within a given distance interval. So, usually students not attending the closest school prefer colleges close to the city center. However, the specifications for which centrality enters the stochastic part show heterogeneity within the population. Dependent on model specification, round 31-34 per cent of the population evaluate C_s positive. If so, schools located at the outskirts yield higher utility (high values of C_s, see Table 3.3 and Figure 3.3). This is because students who prefer the closest school and who are located at the outskirts gain highest utility from the closest "outskirt school".
As expected, all profiles increase utility, particularly the unique profiles (2 and 4) offered by magnet schools. The preferences for the profiles are ordered: The most preferred profile is *languages core* (profile 4) followed by *math/science core* (profile 2) and third *languages* (profile 3). This

order is independent from model specification. The dummy variable, which indicates whether the closest school is a magnet school, shows the expected negative sign and enters significant for all specifications. As assumed the variable compensates unrealistic high utility due to linear distance measure for the students located very close to magnet schools. The coefficient of school size V_s is consistent over all model specifications. It was foreseeable that the sign would be positive, since larger schools obtain scale effects (they offer more extracurricular courses, for example) and the probability that a student's application of enrollment will be rejected is lower than for smaller schools. The sign for authority responsible is positive for MNL specification. In contrast, for the MMNL specifications the sign is always negative. The standard deviation is quite high (2.538 for example). This means that this parameter is highly variable (the 95 per cent interval is [-4.94,3.41], and the 90 per cent interval is [-4.01,2.49]). This makes interpretation difficult to some extent. However, some information is given by this parameter. The mean is negative while 37-39 per cent of all students would evaluate private schools positive. This latent surplus could be conditional on the properties of most of the private schools, like confessional affiliation, elitist student community because of admission test and to some extent specific teaching conditions, which are different from public colleges (ecological subjects for instance). The majority of students regard these properties as repellent. According to this, the choice probability of a private school will be mostly lower than for a public school, especially for those ones far away from students home. But Table 3.3 indicates for the private school (St. Benno) an average commuting distance which is greatly above the overall average. So, students are willing to travel to private schools over a relatively broad geography. In contrast the MNL suggests all students evaluate private schools very positive (1.01) and therefore it neglects that some students dislike private schools. So the MNL does not consider taste variation which is observable using MMNL (see Figure 3.4).

The substitution patterns that are implied by the models can be shown by an easy example. Suppose a private school C offering profile 2 (math/

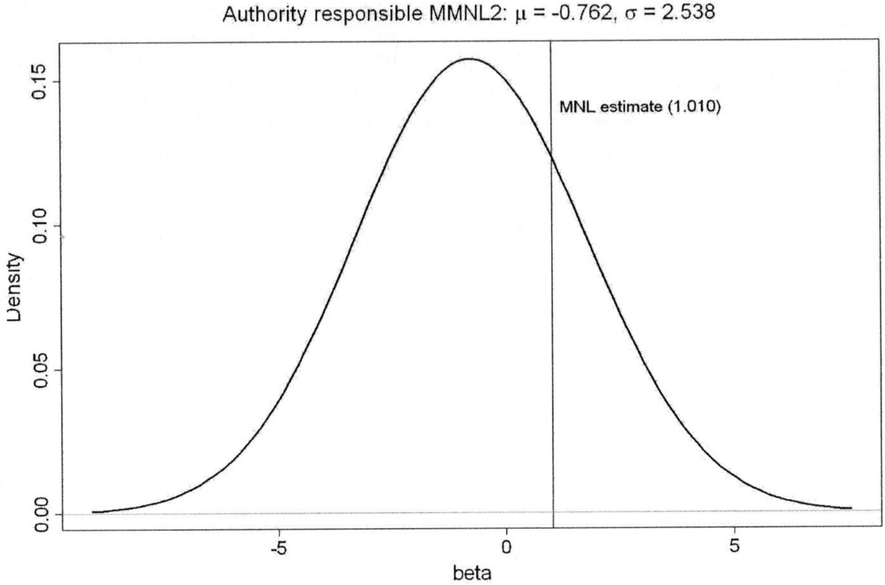

Figure 3.4: Normal Distribution of the Coefficient for the Variable "Authority Responsible" of MMNL2

science core) is introduced to a base situation consisting of

- a private school A offering profile 3 (languages) and
- a public school B offering profile 3 also.

The MNL, because of the IIA property, implies that the new school C will draw proportionately from both schools A and B offering profile 3. In contrast, the mixed logits predict that school C will draw more proportionately from school A than from school B. Using the estimates of MNL the ratio of choice probabilities of A and B is 2.746 for the base situation and the situation with three schools. Using MMNL1 with 10 draws we get a ratio of 1.378 for the base situation and 1.030 for the situation with three schools. If we compute the relative deviation between the two situations for both models this yields 0.523 for A and B using MNL and 0.458 for A and 0.275 for B using the MMNL. Since we want to map substitution patterns in future school network scenarios the MNL is not appropriate here, although it has a better goodness-of-fit. Since all three MMNL differ significantly (see Table 3.4), MMNL2 will be used for simulation of scenarios in the next section, because it has the best explanatory quality. McFadden R^2 is a logistic analogy to R^2 in ordinary least squares (OLS) regression. In general, the McFadden R^2 greater than 0.4 can be interpreted as a very good goodness-of-fit (Backhaus et al. 2003).

3.3 Simulation of School Choice Probabilities

The local school authority of Dresden decided in 2000 to close three colleges in 2008 due to low student numbers. In 2006 the decision has been made to establish three new school locations in 2012 because of increasing student numbers. In this section we utilize MMNL2 to simulate choice probabilities for three different scenarios: In 2004 - the year when the survey was undertaken - there are 20 colleges open (first scenario). In 2008, the second scenario, the colleges Joseph-Haydn, Fritz-Löffler and Großzschachwitz will be closed. This yields a reduced choice set (17 colleges) for simulation. In the third scenario (2012) the new colleges Bühlau, Seidnitz and Bürgerwiese will be opened (see Figure 3.3). So the choice set consists of 20 available schools in 2012. Although small scale data (block level) on student numbers and commuting distances are available, the choice probabilities for each scenario are simulated on district level for visualization purposes. No data is available for the surrounding districts of Dresden. Admittedly, there are no colleges located in the surrounding districts close to the city limit. So the simulated choice probabilities are expected to be realistic, since no colleges outside Dresden would attract students from within the city limits. We employ Biosim 1.4 for simulation (Bierlaire 2005). The number of draws is set to 1000 to derogate variance.

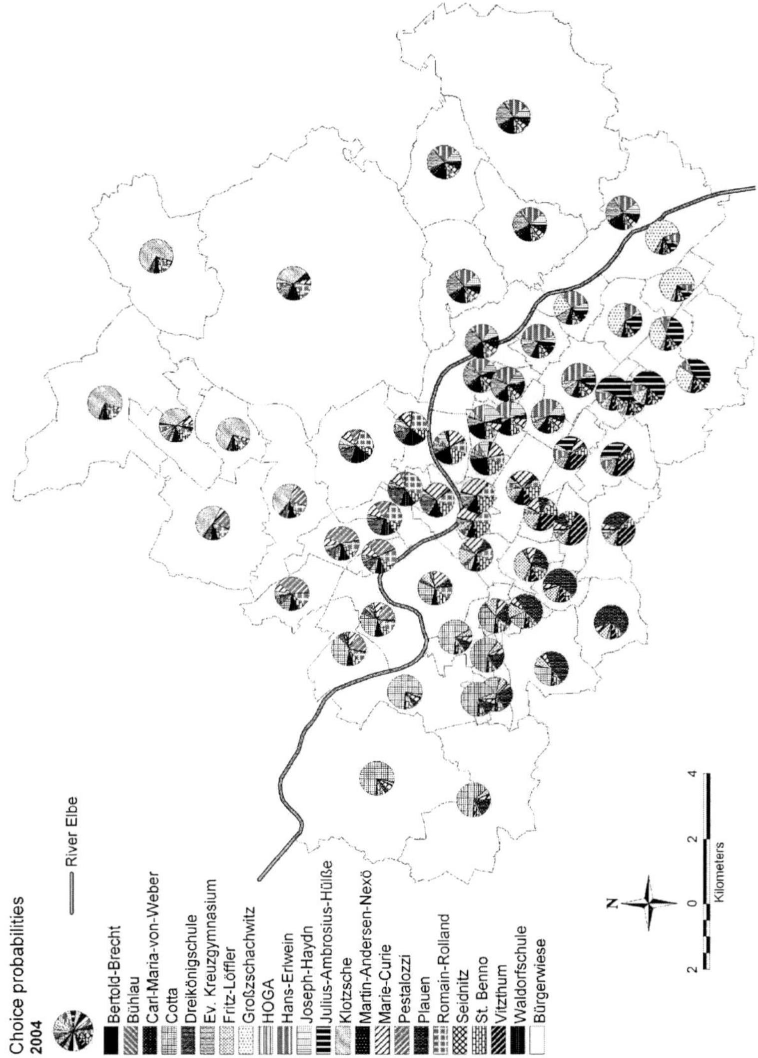

Figure 3.5: Scenario 1 - 2004

Figure 3.5 displays the simulated choice probabilities for 2004. Expectedly, districts located at the outskirts show high choice probabilities for the closest college due to the generally long commuting distances to other college locations. This is Klotzsche College in the north and Cotta College in the west for instance. Districts located closer to the city center show a more diversified choice probability pattern, because of significantly more schools within a small distance interval. It is noteworthy, that the districts in the south-east (north of the river Elbe) have a diversified choice pattern, too. Although they are located at the outskirts. This is conditional on the lack of river crossings within this area and no college is located within these districts. Due to missing ferries or bridges, the river has a spatially strong separating impact in this area. Comparing the choice probabilities of the two most south-eastern districts both sides of the river, this impact becomes apparent. The closest bridge is located approximately 10 kilometers away from the south-eastern edge following the river toward the city center. At this point exists access to a spatial cluster of colleges (for example: Martin-Andersen-Nexö, Carl-Maria-von-Weber, Ev. Kreuzgymnasium, see Figure 3.3). This spatial characteristic leads to the observed diversified choice probabilities.
Besides these spatial distinctions of choice probabilities there are patterns observable, which are more alternative specific. For many districts there is a relatively high choice probability of Romain-Rolland College, which is a magnet school. This college is very attractive for many students throughout the city due to the combination of profile 3 and 4. The location close to the city center implies good accessibility, especially by public transport. There is a particularly high choice probability for this college from the northern outskirts. The dummy variable which identifies the closest school as a magnet school yields the desired effect: the districts close to Romain-Rolland College do not show unrealistic high probabilities.
The most noticeable difference between scenario 1 and 2 (see Figure 3.6) is the increase in choice probability of Julius-Ambrosius-Hülße College for the districts in the south-east conditional on the closure of

Großzschachwitz College. A comparable phenomenon can be seen in the south-east/central districts. Here, the closure of Fritz-Löffler College leads to increased choice probabilities of Plauen and Vitzthum College. The shift in choice probability due to the closing of Joseph-Haydn College is not really discernible compared to the one mentioned above. The location of Joseph-Haydn within a spatial cluster of several other colleges offering the same profile (Carl-Maria-von-Weber, Bertold-Brecht, for example) leads to only marginal changes in probabilities. The opening of Bühlau College in 2012 (see Figure 3.7) results in a dramatic shift of probabilities in the eastern districts north of the river Elbe. In 2012 these districts show the same pattern like districts in the north and south-west outskirts. The probability of Bürgerwiese College in the (south-) central districts and the probability of Seidnitz College (south of river Elbe) are less prominent but still noticeable. All together, the most significant changes in choice probabilities occur in the south-east of Dresden, particular due to the closure of Großzschachwitz College and the opening of Seidnitz and Bühlau College.

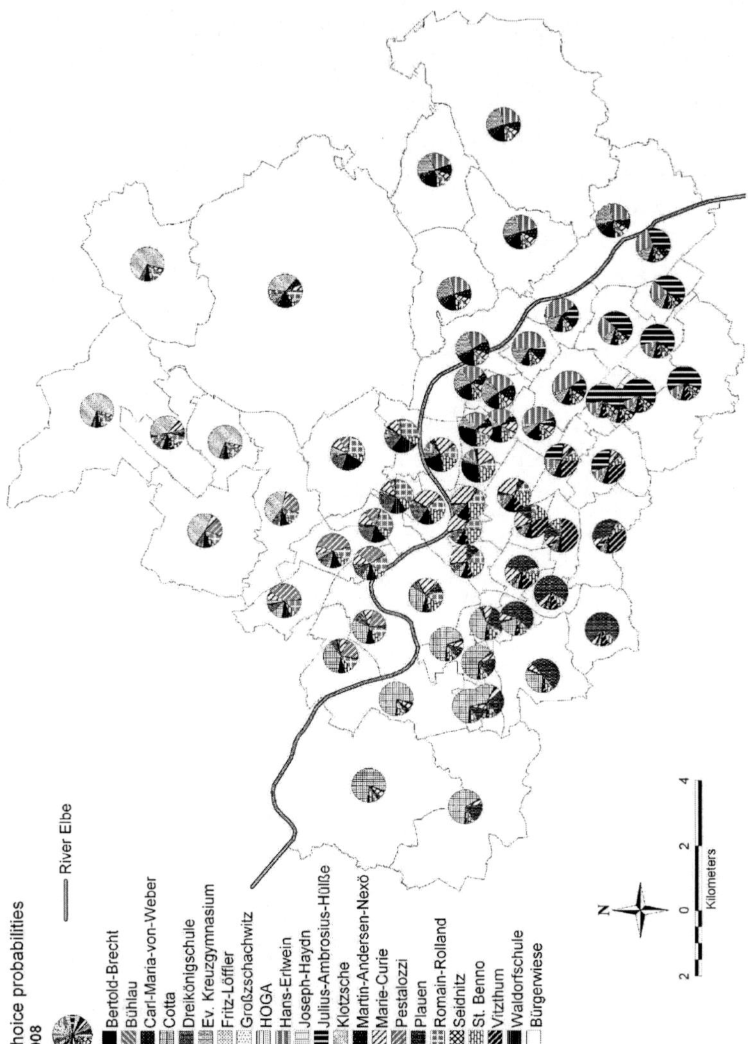

Figure 3.6: Scenario 2 - 2008

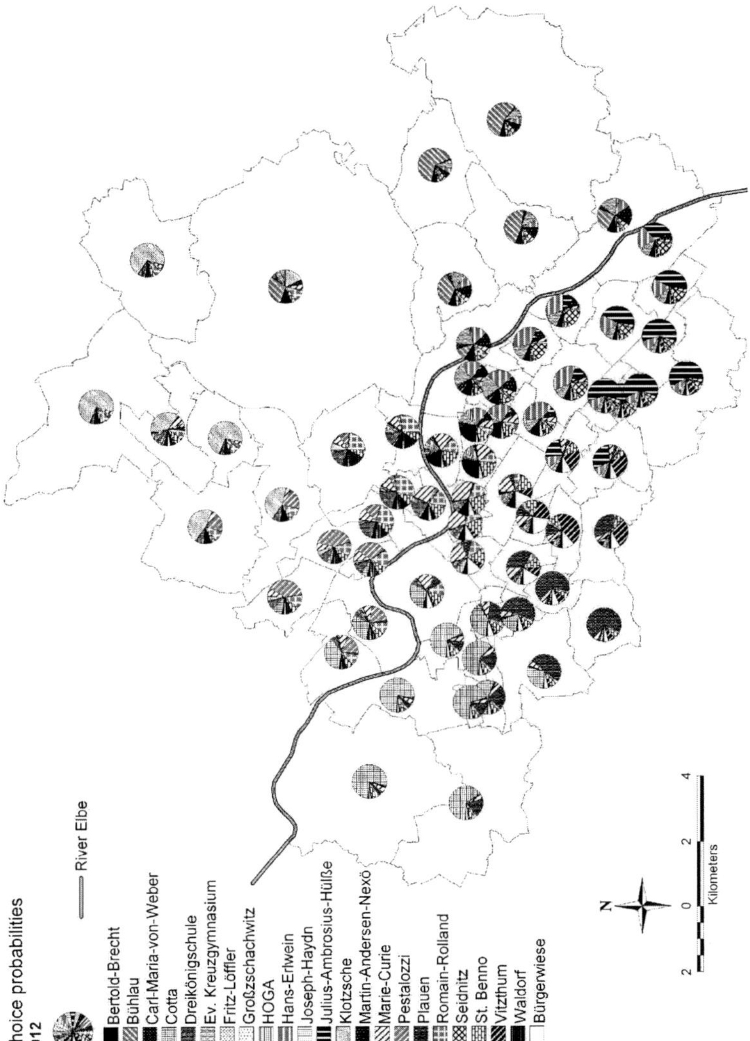

Figure 3.7: Scenario 3 - 2012

Figure 3.8 displays the share of students for each college in the years 2004, 2008 and 2012. The shares have to be interpreted as the potential of students, since capacity is not considered here. Schools located in the city center exhibit a relatively high share of students due to C_s and school size (Marie-Curie) or combination of profiles (St. Benno, Romain-Rolland) as well as the relative large number of students located near the city center. For some colleges (Klotzsche, Pestalozzi, for example) the changing choice sets have a small impact, because they are located far away from areas affected by school closures or openings. It is striking, that the shares of the colleges Julius-Ambrosius-Hülße, Hans-Erlwein, Vitzthum and Plauen are very fluctuant. The net gain or loss of each college for the time span 2004-2012 is given by Figure 3.9. The colleges Julius-Ambrosius-Hülße, Plauen, Cotta and Vitzthum gain student shares, because of the school closures in their proximate surroundings. The increase of student shares for Marie-Curie and St. Benno is due to their attractive location mainly, while the gain of shares for Romain-Rolland College is due to its combination of profiles. Colleges Cotta and Martin-Andersen-Nexö have gained shares although colleges Bürgerwiese and Bühlau have been opened within their surrounding. This is due to size (Cotta) and profiles (Martin-Andersen-Nexö). In contrast to this, the colleges Ev. Kreuzgymnasium and Carl-Maria-von-Weber loose more students to the new colleges than they have gained from the closed ones. These two schools are located in the center close to the three new schools.

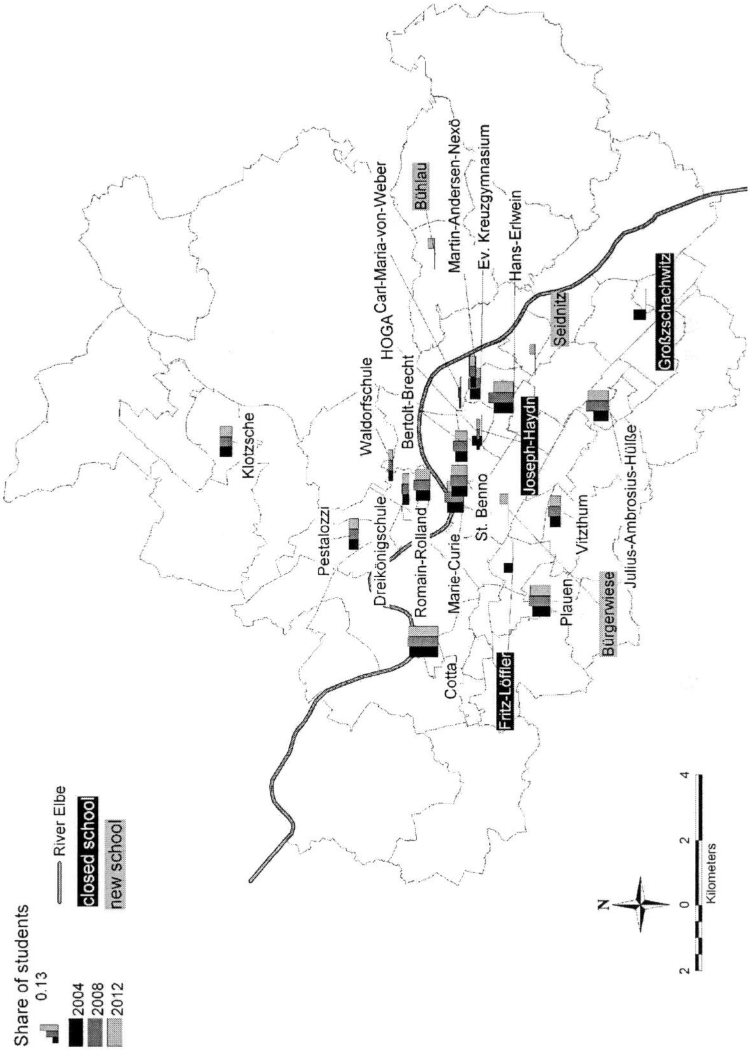

Figure 3.8: Share of Attending Students for the Years 2004, 2008 and 2012

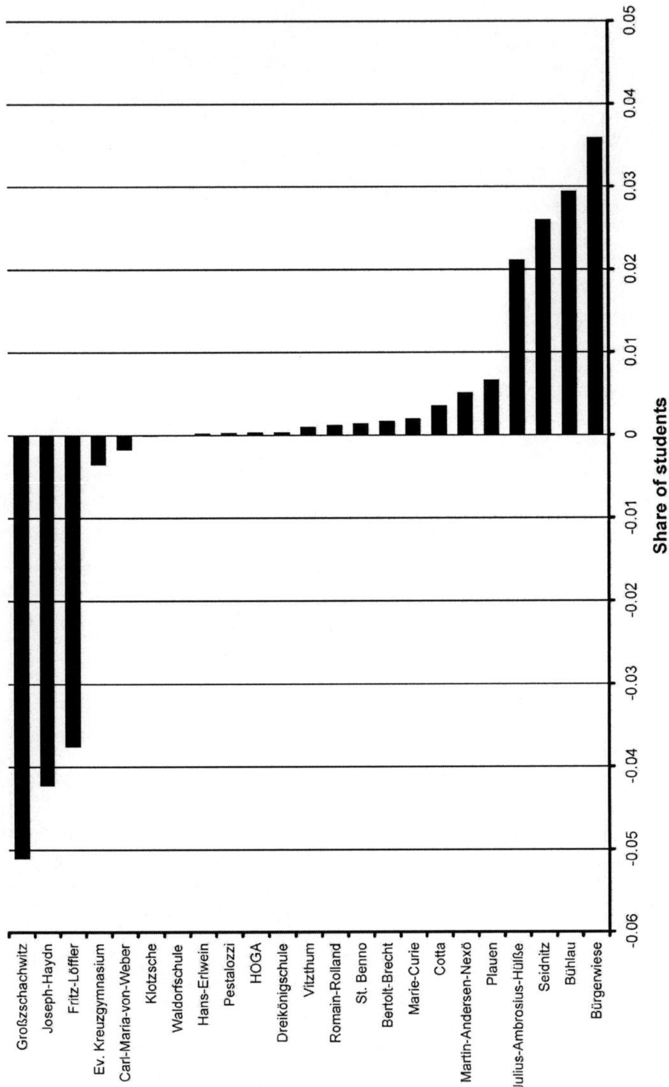

Figure 3.9: Net Gain and Loss in Student Share 2004 - 2012

3.4 Outlook

Our analysis reveals that the size of a school and its service in terms of offered profiles as well as spatial patterns (here: centrality) influence school choice, besides the known attributes as stated in the literature like racial composition, academic quality, sex, proximity and authority responsible (see sections 3.1 and 3.2). Spatial factors have a great impact on school choice. Educational contents are more subordinate while still important. Moreover, our results suggest that the MMNL is appropriate to regard flexible substitution patterns between schools. The surrounding districts of closed or opened school locations show plainest changes in choice probabilities. Nevertheless, the scenario simulation detects changes in probability for more distant districts. The net gain and loss of student share for each college is associated with these results. Even colleges located far away from closed or opened colleges show a distinct variation of student share over time. Thus, an approach which assumes each student chooses the closest school location is not appropriate for dynamic school network planning. Future research determining school choice behavior should incorporate individual-specific data, particularly spatial dependencies of individuals and alternatives. A survey which covers stated as well as revealed preferences would give new insights into school choice behavior. To get more realistic information on student shares one should consider the surrounding regions of Dresden for modelling as well.

4 The Journey-To-School

We do not assume substitution effects between the commuting modes walking, cycling, public transport and car or motorcycle on the journey-to-school. Moreover, we do not expect heterogeneity within the student population concerning the commuting modes. Therefore we use a MNL in order to analyse the travel-to-school mode choice. Section 4.1 covers the choice modelling of the commuting modes. In Section 4.2 we present an example of school closure and modal shift for the City of Dresden. Some final remarks can be found in Section 4.3.

4.1 Travel-to-School Mode Choice Modelling

Regarding the travel-to-school mode choice, the mode is a categorical variable. We suggest a student chooses the commuting mode with the highest utility. So we revert to a MNL since it is based on utility theory and appropriate for categorical data analysis. The logit approach has been widely used in fields of transport modelling (Koppelmann and Sethi 2000, Bhat 2000). The modeler assumes the utility F_{km} of a commuting mode m (walking, cycling, public transport and car/motorcycle) to a student k, and includes a deterministic component v_{km} and an additive random component ϵ_{km}:

$$F_{km} \quad = \quad v_{km} + \epsilon_{km} \tag{4.1}$$

Here, the deterministic component of the utility function is linear in parameters. Assuming that the random component, which represents errors in the modeler's ability to represent all the elements that influence the utility of a commuting mode to a student, is extreme value IID across students and commuting modes, the MNL is as follows:

$$\rho_{km} = \frac{\exp(\upsilon_{km})}{\sum_{m'=1}^{M} \exp(\upsilon_{km'})} \tag{4.2}$$

where ρ_{km} is the probability that commuting mode m is chosen by student k and M is the set of different modes. The closed form of the MNL makes it straightforward to estimate, interpret and use. Detailed work on theory, shortcomings and some applications can be found in McFadden 1973, Ben-Akiva and Lerman 1985, Bhat 1997, Koppelmann and Sethi 2000, Greene 2003. Recent studies concerning travel-to-school mode choice modelling utilizing MNL focus on urban form, built environment and distance (McMillan 2003, Black et al. 2004) as well as travel-time (Woodside et al. 2002, Ewing et al. 2004, McDonald 2005). Ewing et al. (2005) and de Boer (2005) focus on the coherences between travel-to-school mode choice and school location, safety, and vehicle emission.

4.1.1 Data for Travel-to-School Mode Choice Modelling

Specifying the students' utility we revert to the variables age, distance, car availability, season or weather respectively. Distance is a continuous variable measured in kilometers. Car availability (all time / not all time) and weather (fair/bad) are dummy variables. Car availability means whether the student has the possibility to commute to school by car. This includes the possibility of the student to be a passenger while the mother, for instance, drives the car. Car availability equals one, if the student has the possibility to commute to school by car every day.

We consider a small number variables for forecasting purposes and for ease of interpretation. Table 4.1 and Figure 4.1 display an aggregated overview of the survey data set. More information about the data is given in Section 2.1. It is noteable that the average distance of public transport and car/motorcycle increases in summer while the absolute number of students decreases. We suggest that this is because, in summer (or fair weather), those students able to take the bus or car but who cannot switch to walking or cycling because the travelling distance is too great. In winter (or bad/cold weather), there are some students taking the bus or car for reasons of convenience - to avoid walking in the rain for instance - although the distance to school would be acceptable for cycling or walking. With regard to cycling the slight increase in average distance in summer is related to the strong increase in the number of students choosing to cycle. Some of these additional students who are cycling in the summer show longer distances (using public transportation or car in winter). We expect that the decrease in average distance for walking in winter is conditional on students who switch from cycling in summer (due to distance) to walking in winter due to weather conditions. For example, they avoid cycling in the winter in case of snowfall.

4.1.2 Model Results and Interpretation

The results of the maximum likelihood estimation are shown in Table 4.2.[1] Pre-examinations show that age is not significant for the mode public transport and this variable has been removed from the model. There are 4650 college students within our data set. The sample size for estimation is doubled to 9300 because each student is counted once for summer and once for winter. All variables are significant at the 1 per cent level. Table 4.2 shows that on average 81 per cent of all cases are correctly predicted by our model. This in conjunction with the high McFadden R^2 suggests the model has good explanatory qualities. Compared to the mode car/motorcycle all other transport modes have

[1] The author provides the data and the source code on request.

variable	mean	standard deviation	n cases
distance [km]	3.941	3.566	4644
distance - walking (summer)	0.710	0.408	845
distance - walking (winter)	0.859	0.526	1010
distance - cycling (summer)	2.364	1.307	1130
distance - cycling (winter)	2.022	0.988	349
distance - public transportation (summer)	5.390	3.111	2390
distance - public transportation (winter)	4.943	3.034	2838
distance - car/motorcycle (summer)	7.693	6.910	279
distance - car/motorcycle (winter)	6.036	6.104	447
		winter	summer
		n cases	n cases
car availability - all time		391	502
car availability - not all time		4253	4142
season (summer/winter)		4644	4644

Table 4.1: Mean Distances for Each Transport Mode and Absolute Car Availability

a higher utility. If we ignore other influencing factors, walking is the most preferred mode (10.774). Taking into account the other variables, it is obvious that the utility of car/motorcycle will increase in relation to the three other modes. Although some information is provided by the coefficients themselves, the interpretation of the choice probabilities is more revealing. Figure 4.2 shows the travel-to-school mode choice probabilities.

Walking is the most important mode for short distances (up to 1 km) regardless of car availability and weather (a). Concerning cycling (b), weather and distance have a strong influence on associated probability. Whereas students with car availability, switch from bike to car at shorter distances, those with no car available switch from bike to public transportation. To discriminate between the modes public transport (c) and car/motorcycle (d) the stated car availability is the most important factor. The gap between summer and winter in both modes within the range of 1 to 3 kilometers is related to the reduced probability to commute to school by bike in winter. Mostly, distance influences the

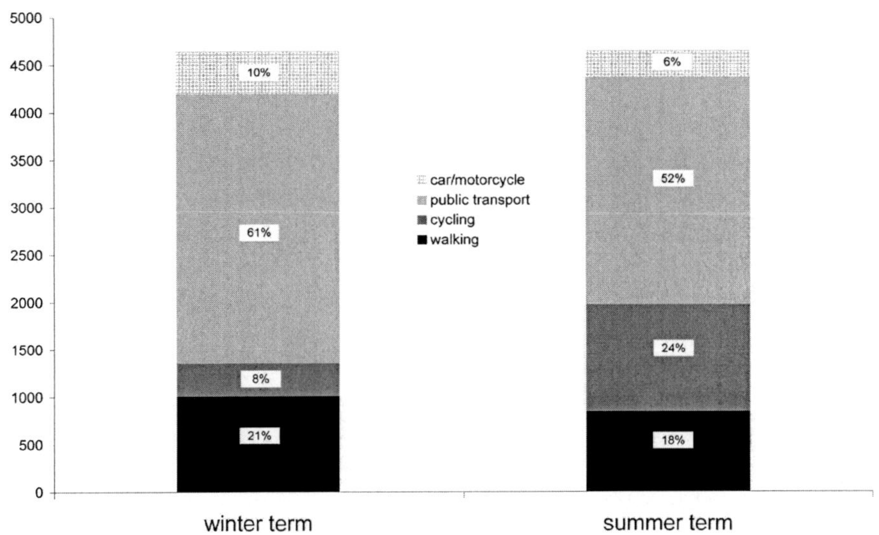

Figure 4.1: Modal Split for College Students in Dresden

decision to go by bike or walk on the one hand and to use public transportation or car/motorcycle on the other hand. For several reasons it is recommendable to avoid a high proportion of students choosing transport modes other than walking or cycling. Obviously, there are higher costs related to motorized transport modes. Moreover, walking and cycling are more activity related and thus better for students' health than motorized transport modes. A large percentage of students using public transport or car/motorcycle yields a negative impact on the environment due to emission (noise/pollution). In the next section we will discuss these issues in more depth.

	MNL Coeff.	Wald	% correctly predicted
walking			88
absolute term	10.774	2119.891	
distance	-4.376	1369.57	
winter/bad weather	-0.591	13.024	
car avail. (all time)	-5.279	489.696	
cycling			42
absolute term	6.57	1196.853	
distance	-0.904	748.843	
winter/bad weather	-2.081	200.51	
car avail. (all time)	-4.772	675.716	
public transport			89
absolute term	4.477	686.946	
distance	-0.052	8.796	
winter/bad weather	-0.489	1510.892	
car avail. (all time)	-0.553	13.111	
car/motorcycle			81
average			81
Number of obs.			9300
2 log likelihood			-9584
McFadden-R^2			0.63

all variables are significant at the 1 per cent level
reference category: car/motorcycle

Table 4.2: Coefficients and Goodness-Of-Fit of the Travel-To-School Mode Choice Model

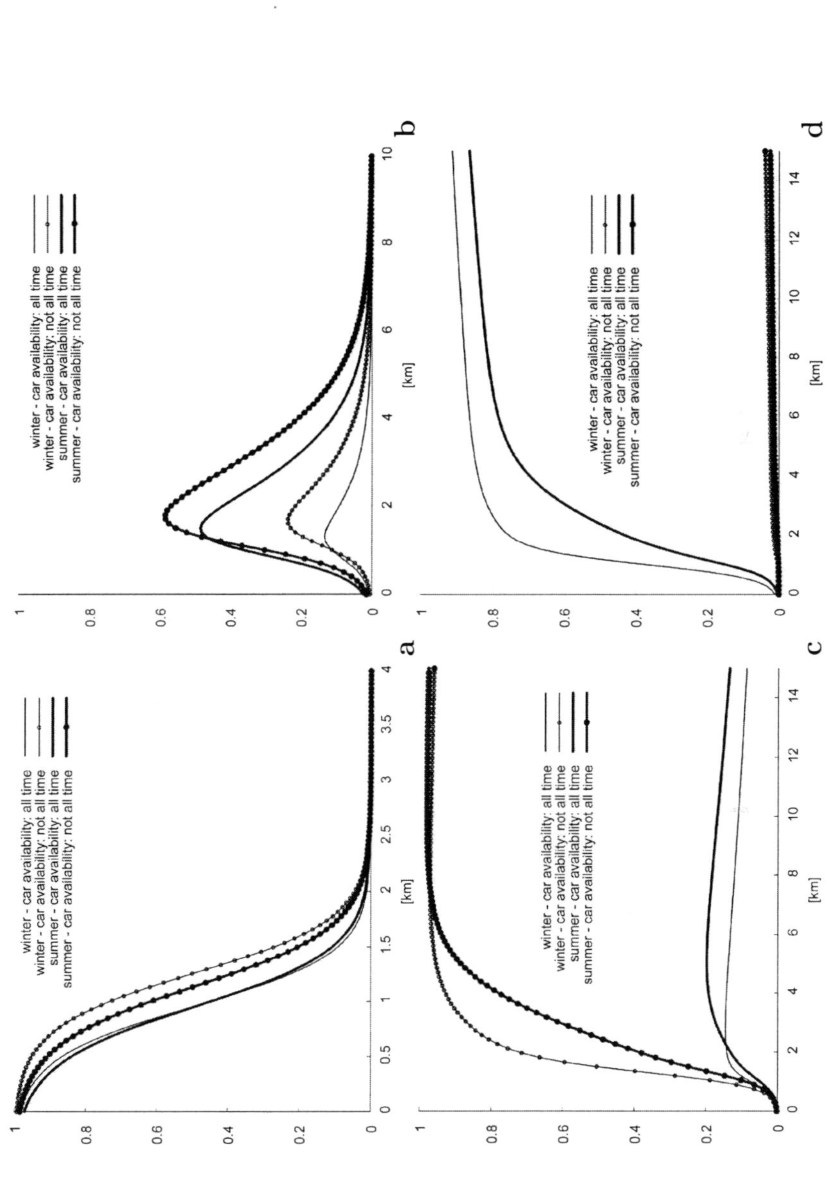

Figure 4.2: Mode Specific Choice Probabilities (a: Walking, b: Cycling, c: Public Transport, d: Car/Motorcycle)

4.2 Modal Shift and School Closure - An Example

Under-utilization usually forces authorities to close schools. This is often justified by economic aspects. In this section we analyze whether there is economic and ecological evidence emerging from a modal-shift, which could justify keeping selected under-utilized schools open.

4.2.1 Modal Shift Due to School Closure

In the year 2000, the school authorities in Saxony decided to close several school locations in Dresden due to declining enrollment in the 1990s. One of them is Großzschachwitz College which will be closed in summer 2008. Because of this, affected students have to attend alternative schools. Here, we analyze the travel-to-school mode choice probabilities and the related consequences.
For the reasons outlined in the introduction (Section 1.3) we do not use the MMNL specified in Section 3.2 in order to determine the school choice probabilities and the number of students attending available schools. Instead we refer to the coherences of Section 3.1.2. According to Table 3.1, there are 68 per cent of the students located in zone 1 attending Großzschachwitz College. In 2004, there are overall 467 students enrolled on Großzschachwitz College. Hence, 318 students of Großzschachwitz College are located in zone 1. The total number of college students in zone 1 from Großzschachwitz College is 403 in 2004. Thus, 79 per cent of all college students located in a block of zone 1 of Großzschachwitz College attend this college in 2004. Based on this, we assume that 79 per cent of the students located in zone 1 enroll on the closest college available (see Section 3.1). We apply the MNL specified in Section 4.1 and yield the number of students[1] choosing a given commuting mode for

[1]We have computed the average probability due to summer and winter term.

the years 2004 and 2008 (see Figure 4.3 and 4.4). In both cases we just consider the 79 per cent of the students located in zone 1 who attend the closest college, which is Großzschachwitz College in 2004 and Julius-Ambrosius-Hülße College in 2008. Three main patterns are evident:

1. Usually there is no possibility for most of the students to commute to school by car (see Figure 4.1 and Table 4.1). So, in both scenarios there is only a small number of students commuting to school by car or motorcycle.

2. Although no strong difference in the absolute number of students commuting to school by bike can be observed, one can identify a difference in the spatial pattern: the spatial center of gravity of students who commute to school by bike shifts toward the location of Großzschachwitz College.

3. Most obviously there is a strong increase in the use of public transport while far less students walk to school in 2008. Note that the number of students in zone 1 of Großzschachwitz increases from 2004 to 2008.

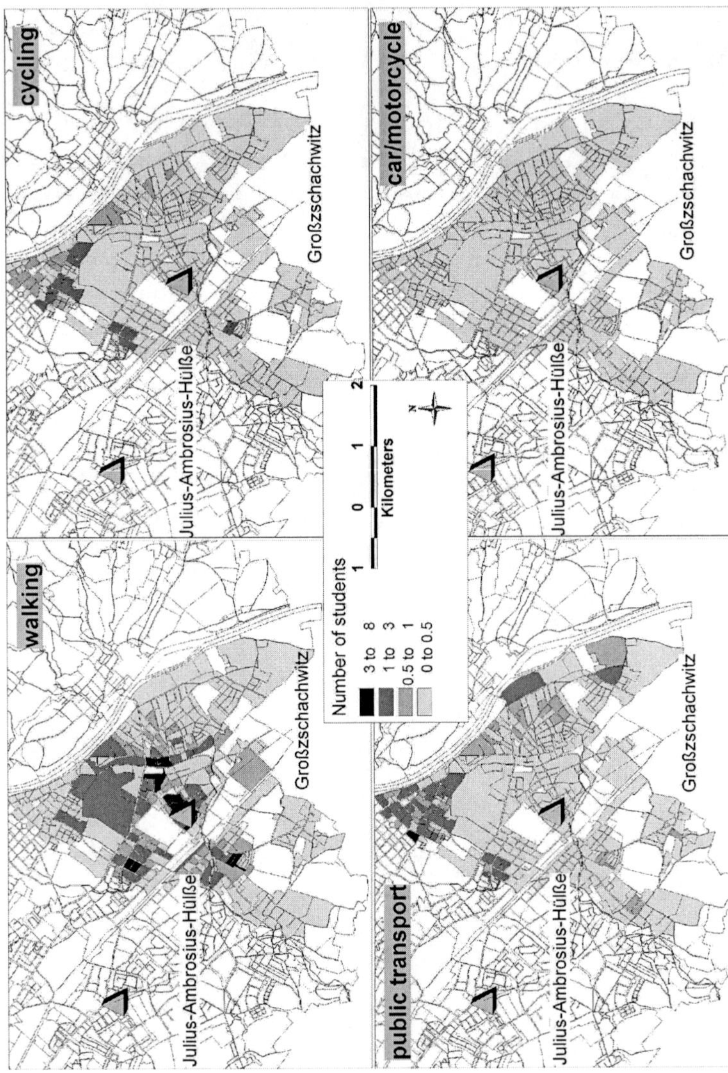

Figure 4.3: Number of Students Walking, Cycling, using Public Transportation or Car/Motorcycle in 2004 (Before the Closure of Großzschachwitz College)

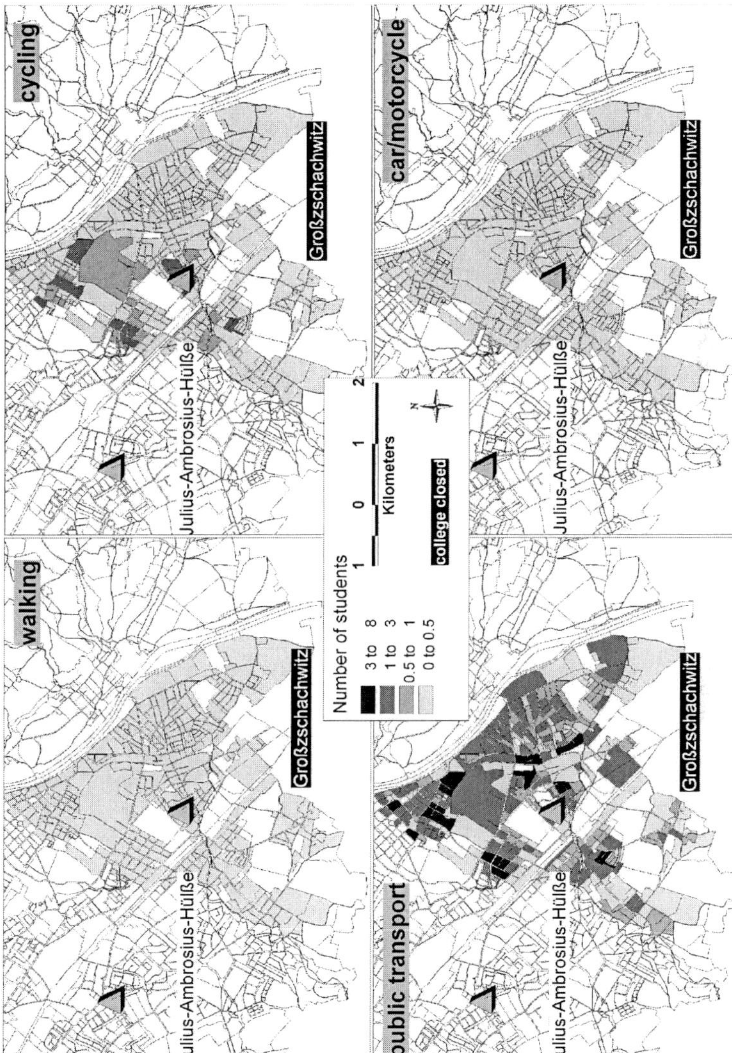

Figure 4.4: Number of Students Walking, Cycling, using Public Transportation or Car/Motorcycle in 2008 (**After** the Closure of Großzschachwitz College)

4.2.2 Quantification of Modal-Shift

Here we try to quantify the consequences of the modal-shift due to a school closure. We focus on the transport sector, so we ignore costs related to the school location like maintenance and rent as well as external diseconomies of closing a school like the loss of local neighborhood community (shops and services that depend on local schools are forced to close for instance). We are aware of the difficulties associated with quantifying the modal-shift by costs since these costs are not always easy to determine - particularly external diseconomies. For convenience we do not discuss the different cost figures stated in the literature (Infras/IWW 2004, Planco Consulting GmbH 1993 and Bickel and Friedrich 1995 for example). Here we use the costs given in Table 4.3.
All figures are costs per trip. As plain costs we consider the average usage and consumption costs for cars and bicycles covering fuel usage, insurance as well as purchase and maintenance costs. The bus or tram fare reflects the costs of one trip using a seasonal ticket. We carefully assume that a student makes 2.2 trips to school per day[1]. There are usually 200 days of school per year in Saxony. The travel-times are derived from the distance matrix and the assumed average speeds (see Chapter 2). For public transport travel-times we use a travel-time matrix based on the bus and tram line network of 2004. We do not explicitly consider congestion costs because we assume these are included in the value of travel-time, pollution and noise costs. Furthermore, there arise costs due to decreased physical activity which is related to the motorized commuting modes. An increase in the number of students commuting to school by car or bus yields increased levels of obesity, type 2 diabetes, heart disease etc. Unfortunately, we do not have information about the relationship of ill students and students choosing motorized transport modes, nor do we have costs figures available based on diseases.

[1]One trip to school in the morning and one trip back home at midday per school day. At some days there are additional trips necessary in the afternoon (sports for example).

name	costs (€)	unit	source
plain costs			
cycling	.005	student-km	assumed
bus/tram (fare)	1	student	VVO (2007)
car/motorcycle	.165	student-km	FGSV (2002)
value of travel-time			
walking	.03	student-minute	Baum et al. (1998)
cycling	.035	student-minute	assumed
public transport	.04	student-minute	Axhausen et al. (2001)
car/motorcycle	.065	student-minute	Axhausen et al. (2001)
accident			Baum et al. (1998)
public transport	.28	student-km	
car/motorcycle	1.64	student-km	
noise			Baum et al. (1998)
public transport	.005	student-km	
car/motorcycle	.006	student-km	
pollution			Baum et al. (1998)
public transport	.007	student-km	
car/motorcycle	.015	student-km	

Note that all figures are costs per trip

Table 4.3: Transport Costs Figures

Figures 4.5 and 4.6 present the mode specific transport costs assigned to the location of the originator (student) for 2004 and 2008. In 2004 the walking costs are due to the accumulated value of travel-time of many students walking to school with distances up to 1.5 km. Due to longer distances the number of students walking is very low in 2008 - and, therefore, so are the walking costs. There is an increase in cycling costs observable, particularly within proximity of Großzschachwitz College. This is reasonable since there is a remarkable increase in student numbers in this area. Moreover, more students go by bike due to longer commuting distances. The increased number of students is a cogent reason for the increase in public transport costs. However, the majority in cost variation is because of the modal-shift due to longer commuting distances caused by the closure of Großzschachwitz College. There is a remarkable increase in students commuting by public transport, in particular within proximity to Großzschachwitz College. Because of the low level of car availability the mode car and its costs are negligible. In absolute numbers the transport costs rise from nearly 80.000 € in 2004 to more than 200.000 € in 2008 (increase of 150 per cent). This increase is mainly due to public transport costs, and focuses spatially on the proximate area (radius of 1 km) of Großzschachwitz college (see Figure 4.7). Assuming realistic location costs of a college of more than 1 million € per year (see Section 6.1), the increase in transport costs does not satisfy the decision to keep a college running which suffers from under-utilization. This will probably hold true even if one considers all students enrolled on Großzschachwitz College and additional external diseconomies (health, loss of community).

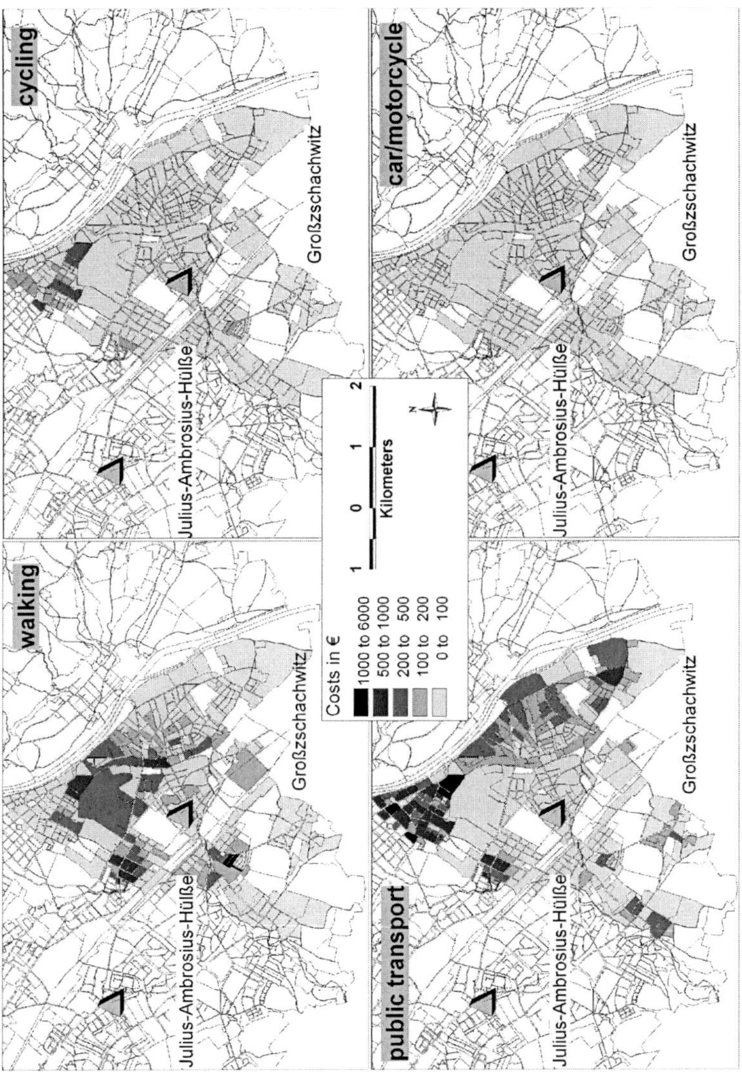

Figure 4.5: Transport Costs per Mode in 2004

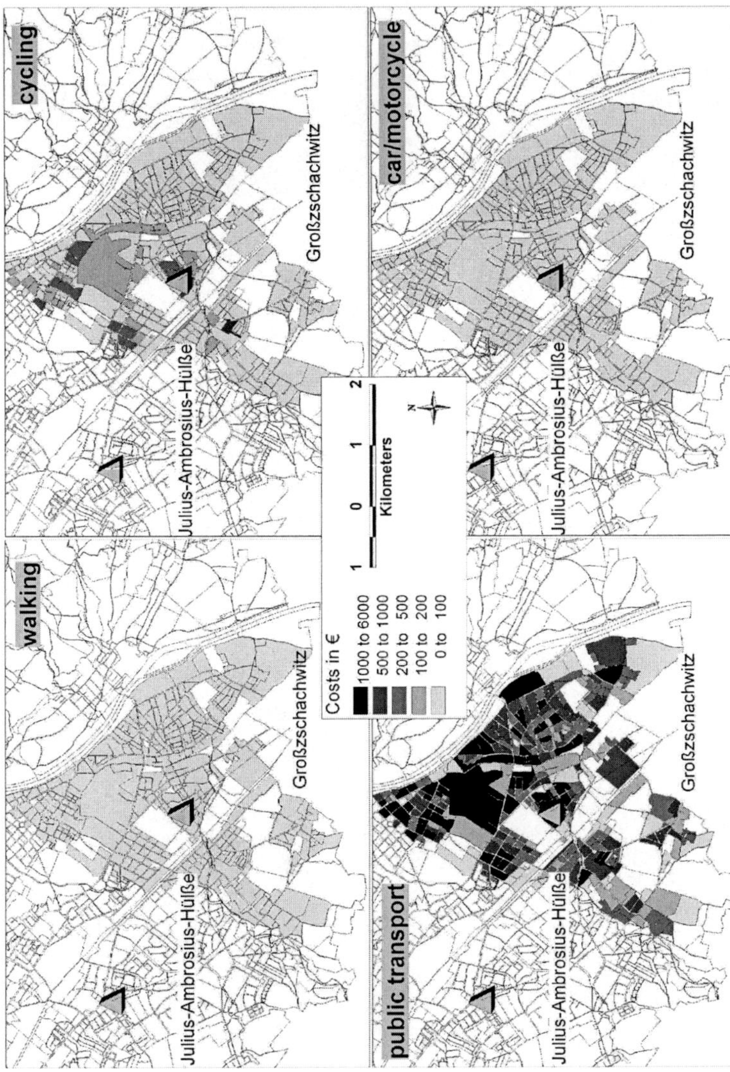

Figure 4.6: Transport Costs per Mode in 2008

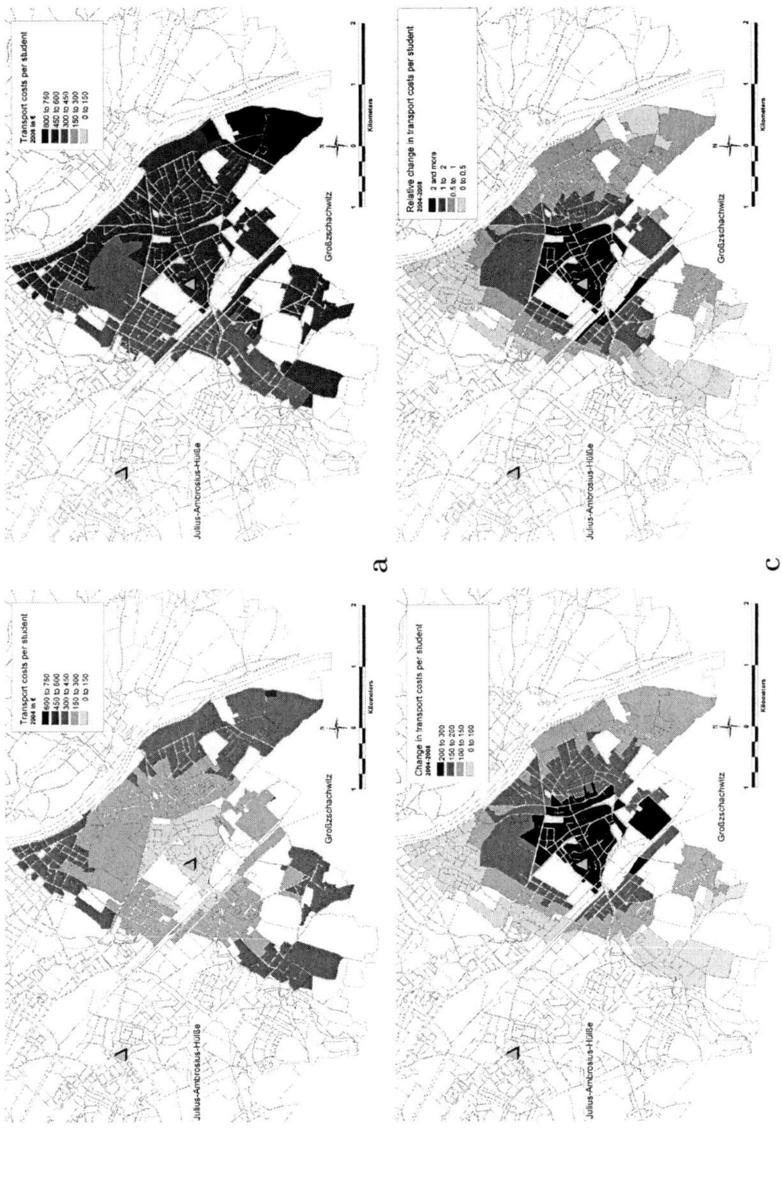

Figure 4.7: Zone 1 of Großzschachwitz College: Absolute (c) and Relative (d) Change in Transport Costs per Student from 2004 (a) to 2008 (b)

4.3 Outlook

Our findings concerning the journey-to-school are consistent with the literature in the field. Moreover, the findings generate robust empirical evidence due to the extent of the sample. By way of an example we attempted to quantify the costs of a modal-shift due to a school closure. Although the increase in transport costs is remarkable this is not a substantial reason - from an economic point of view - against a school closure within an urbanized area. If we consider other factors, such as the health of the students or ecological aspects, the costs of a modal-shift becomes more onerous.

A simultaneous school and travel-to-school mode choice model should give more insight into the coherences of educational service and accepted travel-time for different profiles, for instance. An interesting question is whether compensation payments (see Section 6.1) have an impact on school and travel-to-school mode choice. A nested logit model could be an appropriate method for such an analysis.

5 School Network Planning

In this chapter an approach is demonstrated which minimizes the location and transport costs with respect to school capacity and utility maximization of students. A large scale optimization approach is presented. The remainder of this chapter is organized as follows: In Section 5.1 we discuss existing problems in the field. Section 5.2 contains the construction of the models. The chapter finishes with some final remarks and comments on possible future research.

5.1 A History of School Network Planning Problems

For the planning of school locations we categorize the different approaches found in the literature. We define the *dynamic school district planning problem* (DSDPP) as the assignment of demand points to schools with (fixed) demand and/or capacity over a given planning horizon. The school district planning problem (SDPP) is a special case of the more general DSDPP because only one single period is considered. The generic districting problem and the school boundary stability problem over time (SBS) are special cases of the DSDPP. If there are decisions made per period on capacity and/or school location as well as on the assignment or allocation of students, this is named dynamic school location planning or dynamic school network planning. We use the expression *dynamic*

school network planning problem (DSNPP) since it implies the decisions on school location and student allocation simultaneously. Again, the static SNPP is a special case of the more general DSNPP. Table 5.1 gives an overview of existing (D)SDPP and (D)SNPP.
Yeates (1963) first applied a linear programming approach (LP) to determine school districts as catchment areas with capacity constraints and constraints on racial balance. He, as many authors have done in the following, treats the problem as a standard transportation (distribution) problem (Trans). Clarke and Surkis (1968), Koenigsberg (1968) and Heckman and Taylor (1969) have discussed the SDPP in the 1960s - particularly concerning the relationship of racial and capacity constraints. A lot of work has been done in the first half of the 1970s mostly considering single attribute objective functions (Maxfield 1972, Belford and Ratcliff 1972, Franklin and Koenigsberg 1973, Liggett 1973, Stimson and Thompson 1974, Holloway et al. 1975, McDaniel 1975, McKeown and Workman 1976). Trifon and Livnat (1973) have made the attempt to model a DSNPP. They use a LP so the decision whether a school should be opened at a potential site is implicitly given by capacity expansion. Due to poor computational capabilities in the 1960s and the 1970s only LPs have been considered and some of the approaches could not be applied to real world problems. Bahrenberg (1981) was the first to apply the SNPP to a german region. He focuses on elementary schools and does not consider free school choice. More work in the first half of the 1980s has been done by Jennergen and Obel (1980), Bovet (1982) and Bruno and Andersen (1982). Sutcliffe et al. (1984) use a more general goal programming approach for the SDPP. They are the first to consider the complexity of the journey-to-school as well as gender proportions. Greenleaf and Harrison (1987) solved the DSNPP as a mixed integer program (MIP). In the same year Diamond and Wright (1987) published the first MIP for the multi-objective SNPP. The few contributions during the 1990s are by Church mainly (Schoepfle and Church 1991). Compared to our approach, particularly the work of Church and Schoepfle (1993) is of special interest because it allows for free school choice. The main

differences to our approach are:

- they apply a restricted and ordered choice set (three choices),
- their approach is a single period problem and
- they do not minimize costs.

In contrast, we consider full choice sets, a dynamic problem and cost as well as utility or rather preferences. Since the year 2000 there is an increasing number of publications, particular on larger and more complex problems (Lemberg and Church 2000, Antunes and Peeters 2000, Caro et al. 2004, Antunes et al. 2004 and Borges and Antunes 2006). Hanley (2007) presents an approach, combining bus route planning and district resizing in a mixed integer non-linear program (MINLP).
Besides these approaches, there are more contributions to the topic of dynamic school network planning. Henig and Gerchak (1986) incorporate demand uncertainty in a deterministic model for dynamic capacity planning of public urban schools. Using multidimensional scaling techniques, Molinero (1988) explores social consequences of closing schools within a framework of declining enrollments. Furthermore, Schoepfle and Church (1989) have developed a fast, network-based, hybrid heuristic for the assignment of students to schools. Koroglu (1992) has developed a probalistic approach to identify optimal locations for new universities in Turkey. The model by Diamond and Wright (1987) has been reformulated by Church and Murray (1993) to remove the inconsistency of minimizing the deviations in capacity utilization between opened and closed schools. For the SNPP Pizzolato (1994) introduces a heuristic algorithm to solve the p-median problem designed for large weighted graphs. He uses principal component analysis, investigating university attractiveness and socio-economic status of demand points. Antunes and Peeters (2001) demonstrate how the complex dynamic modular capacitated facility location problem (DMCFLP) with application to dynamic school network planning can be solved by simulated annealing.

Study & Year	Objective	Class	Cap.	Minority	Demand units	Free choice	All students assigned	Problem
Yeates (1963)	euclid. dist.	Trans	U/L	U/L	Mile sq.		Y	SDP
Clarke and Surkis (1968)	travel-time	Mintran/LP	U	U/L	census tracts		Y	SDP
Koenigsberg (1968)	road dist.	Trans	U	U/L	CED		Y	SDP
Heckman and Taylor (1969)	euclid./ road dist.	LP	U/L	U/L	census tracts		Y	SDP
Maxfield (1972)	road dist.	Trans	U	U/L	1/4 mile sq.			SDP
Belford and Ratcliff (1972)	euclid. dist.	min cost flow	U/L	U/L	city blocks		Y	SDP
Franklin and Koenigsberg (1973)	euclid. dist.	Trans	U	U	CED		Y	SDP
Liggett (1973)	road dist.	MIP	U/L	U/L	city blocks		Y	SDP
Trifon and Livnat (1973)	cost	LP			resid. sections		Y	DSNP
Stimson and Thompson (1974)	euclid. dist.	LP	U	U/L	tracts		Y	SDP
Holloway et al. (1975)	euclid. dist.	LP	U/L		CED		Y	DSDP
McDaniel (1975)	euclid. dist.	Trans	U/L	U/L	city blocks			SDP
McKeown and Workman (1976)	road dist.	LP	U		bus stops		Y	SDP
Jennergen and Obel (1980)	euclid. dist.	LP	U		census tracts		Y	SDP
Bahrenberg (1981)	dist./service	LP	U/L		cities		Y	SNP
Bovet (1982)	dist.		U		towns		Y	SDP
Bruno and Andersen (1982)	cost	LP	U/L	U/L	grid		Y	SDP
Sutcliffe et al. (1984)	dist.(-decay)	LP/GP	U/(L)	U/L	primary schools		Y	SDP
Greenleaf and Harrison (1987)	cost	MIP	U/L		districts/tracts		Y	DSNP
Diamond and Wright (1987)	mult. obj.	MIP	U/L	U/L	tracts		Y	SNP
Schoepfle and Church (1991)	comm. cost	NFP	U	U/L	tracts		Y	SDP
Church and Schoepfle (1993)	preferences	Knapsack	U/(L)	U/L	students	Y		(SDP)
Lemberg and Church (2000)	dist./reassign	MIP	U	U/L	tracts		Y	DSDP
Antunes and Peeters (2000)	cost	MIP	U/L		centers		Y	DSNP
Caro et al. (2004)	dist.	MIP		U	city blocks		Y	SDP
Antunes et al. (2004)	dist.		U/L		centers		(Y)	SNP
Borges and Antunes (2006)	dist./cap. exp.	MIP	U/L		districts		Y	SNP
Hanley (2007)	bus route & district size	MINLP	U		(districts)			(SDP)

U: Upper capacity/share constraint
L: Lower capacity/share constraint
CED: Census enumeration districts
Y: embodied

Table 5.1: Studies on Dynamic School Network Planning

5.2 A New Dynamic School Network Planning Problem

Our approach minimizes the costs of a school network over a given planning horizon while it guarantees maximum student utility for the given school network. The school choice probabilities described in Chapter 3 will be interpreted as the share of students located at a given demand point attending available schools. However, this leads to two problems: First, there is no capacity constraint within the MMNL. Namely, as many students will attend a certain school as determined by the MMNL. Thereby very attractive schools would be overcrowded. Second, the choice probabilities could not be implemented in the school network planning problem directly, due to the decision of students which school to choose depends on the choice set (available schools) which in turn is a result of the school network planning problem itself. Therefore, a two step large scale optimization approach is required.

First, for each period of the planning horizon the enrollment has to be computed for all feasible combinations of open (available) and closed (non-available) schools with respect to capacity. We name the combination of open and closed schools *scenario*. We have to enumerate all possible scenarios. For a given scenario the MMNL has to be applied in order to determine the share of students attending available schools. The possible deviation between the empirical expected share of students and the proportional enrollment on available schools computed by this approach is minimized. The result of the first step is the enrollment of students on schools according to probabilities and school specific capacity. Note, that we are considering only newly enrolled students ("freshers"). Once the decision has been made which school to enroll at, this decision is permanent for the whole primary or secondary school career. Thus, the total number of students attending a given school is known by multiplying the number of 1st or 5th grade students by the number of grades (see Figure 2.1). Therefore, we model the enrollment decisions and deduce the number of students and the related costs from enrollment numbers.

This approach does not allow students to change between schools due to school closures during the school career. We assume that a consistent environment (classmates and teachers) is very important for educational success. Therefore, changing school location due to closure is not desirable for the planning process.

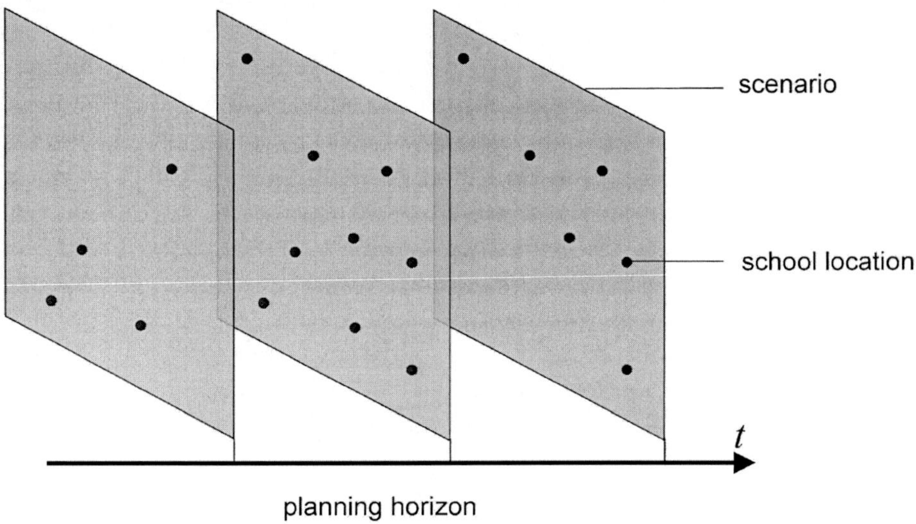

Figure 5.1: Scenario Selection

In a second step for each period one scenario is selected (see Figure 5.1). The costs of each scenario are based on the number of students, location costs and transport costs. Furthermore, this second step provides information on which school is closed or opened in a given period. For example: elementary school A is closed in period 3. So, no enrollment for 1st grade students is allowed on school A as from period 3. However, 1st grade students of periods 2 and 1 still attend this school until the end of their school career. If a school career lasts 4 years, school A will finally

be closed in period 6. At this point in time the last student cohort has graduated and the school could be sold, for example.
The first step (Student Allocation) guarantees students utility maximization and that capacity constraints are met. This is a small and easy to solve problem, but it has to be solved several times. The second step (Scenario Selection) minimizes the school network costs over the planning horizon. This large problem has to be solved only once.

5.2.1 Student Allocation

Let $t \in T$ be the periods, $s \in S$ the schools, $i \in I$ the demand points and $j \in J$ the scenarios. S_j denotes all available schools s embedded in scenario j. J_t is the set of all feasible scenarios j in period t regarding the total number of students in t and the total capacity of all schools $s \in S_j$. The number of students located in demand point i is given by n_i. p_{is} is the empirical expected share of students located in demand point i attending an available school s according to Equation 3.1. α and λ are parameters which have to be specified by the decision maker in order to weight the two parts of the objective (capacity expansion and student rejection).[1] The maximum number of students per school is denoted by G_s which is the product of maximum number of students per class and maximum number of classes per grade.[2] To identify the need for capacity expansion we consider the variable w_s, which is the number of students exceeding the capacity of school s. H_s denotes the accepted amount of students above the capacity. The variable x_{is} represents the proportional enrollment of students located in demand point i on school s. The discrepancies between x_{is} and p_{is} related to capacity are denoted by z_{is}^+ (above) and z_{is}^- (below). The problem of student allocation can be formulated as follows:

[1] For example $\lambda = 1 - \alpha$ with $\alpha \in [0, 1]$.

[2] Since we only consider newly enrolled students, capacity is based on 1st or 5th grade students.

$$\min \quad \alpha \sum_{s \in S_j} (w_s/G_s)^2 + \lambda \sum_{i \in I, s \in S_j} (z_{is}^+ + z_{is}^-)^2 \tag{5.1}$$

subject to

$$
\begin{aligned}
x_{is}/p_{is} - z_{is}^+ + z_{is}^- &= 1 & i \in I, s \in S_j && (5.2)\\
\sum_{i \in I} x_{is} n_i - w_s &\leq G_s & s \in S_j && (5.3)\\
\sum_{s \in S_j} x_{is} &= 1 & i \in I && (5.4)\\
x_{is} &\geq 0 & i \in I, s \in S_j && (5.5)\\
z_{is}^+, z_{is}^- &\geq 0 & i \in I, s \in S_j && (5.6)\\
w_s &\in [0, H_s] & s \in S_j && (5.7)
\end{aligned}
$$

The first part of the objective (5.1) minimizes the weighted number of students exceeding capacity ("capacity expansion"). The second part minimizes the deviation from the empirical expected student shares. The formulation of this quadratic constrained program (QCP) can be construed to some extent within a least squares regression motivation. Constraints (5.2) determine the enrollment and the relative deviation from the empirical expected student shares. Inequalities (5.3) in interaction with (5.7) state that the capacity of school s must not be exceeded more than given by H_s. For each demand point the share of students has to sum to one over all schools available (5.4). The domains are given by (5.5 to 5.7). Note that the problem has to be solved for each combination of $t \in T$ and $j \in J_t$. If no solution exists for $j \in J_t$ then remove j from J_t.

5.2.2 Scenario Selection

Let J_{st} be the subset of all scenarios $j \in J_t$ which contain school s. Moreover, the costs of each scenario $j \in J_t$ for period t are given by c_{tj}. This includes:

(a) the fix location costs of each available school,

(b) the variable costs of each available school and

(c) the transport costs for all students.

The fix location costs (a) are costs related to rent, maintenance and so forth. The variable costs (b) as an element of the scenario costs reflect the costs of 1st or 5th grade students for the whole primary or secondary school career. In other words, we assign the total costs of a student occurring during primary or secondary school career to the year of enrollment. These costs include the penalty costs due to violating minimum and maximum capacity. The transport costs (c) for each scenario $j \in J_t$ and period t depends on the commuting mode chosen for the journey-to-school. Since the distribution of students between demand point i and schools s is known for each period t and scenario $j \in J_t$, it is straightforward to compute the share of students choosing a certain commuting mode by a MNL. Additionally, we have to define the costs of closing or opening a school s in period t by c_{st}^+ and c_{st}^- respectively. Note that all cost parameters are already discounted. The binary variable y_{tj} equals one, if scenario $j \in J_t$ is selected in period t (0, otherwise). x_{st}^+ denotes whether a school s is opened in period t (=1) or not (=0), while x_{st}^- equals one for a school closure (0, otherwise). The DSNPP is as follows:

$$\min \sum_{t \in T, j \in J_t} c_{tj} y_{tj} + \sum_{s \in S, t \in T} (c^+_{st} x^+_{st} + c^-_{st} x^-_{st}) \tag{5.8}$$

subject to

$$\sum_{j \in J_t} y_{tj} = 1 \qquad t \in T \tag{5.9}$$

$$\sum_{j \in J_{st}} y_{tj} - \sum_{j \in J_{s,t-1}} y_{t-1,j} - x^+_{st} + x^-_{st} = 0 \qquad s \in S, t \in T \tag{5.10}$$

$$\sum_{t \in T} (x^+_{st} + x^-_{st}) \leq 1 \qquad s \in S \tag{5.11}$$

$$y_{tj} \in \{0, 1\} \qquad j \in J_t, t \in T \tag{5.12}$$

$$x^+_{st}, x^-_{st} \in \{0, 1\} \qquad s \in S, t \in T \tag{5.13}$$

The objective (5.8) of this MIP minimizes the school network costs over the planning horizon. Constraints (5.9) guarantee that only one scenario is selected per period. (5.10) force a school s to be opened or closed, if two different scenarios are selected in successive periods and only one of the two scenarios contains school s. Constraints (5.11) assure that a school could only be opened or closed once over all periods. A closed school could not be reopened and an opened school could not be closed in following periods. In interaction, (5.10) and (5.11) decrease the number of possible different scenarios selected over the planning horizon. The domains are given by (5.12, 5.13). To view (5.10) in the proper perspective, consider the following cases for a given school s and period t:

(a) In t and $t-1$ the scenarios $j \in J_{st}$ and $j' \in J_{s,t-1}$ are selected. Hence, $y_{tj} = 1$ and $y_{t-1,j'} = 1$ and in combination with the objective function (5.8) x_{st}^+ and x_{st}^- are zero since c_{st}^+,$c_{st}^- > 0$.

(b) In t and $t-1$ a scenario j with $j \notin J_{st}$ and $j \notin J_{s,t-1}$ is selected. Hence, $\sum_{j \in J_{st}} y_{tj}$ and $\sum_{j \in J_{s,t-1}} y_{t-1,j}$ do not exist and therefore x_{st}^+ and x_{st}^- have to be zero since the corresponding coefficients of the objective function (5.8) c_{st}^+,$c_{st}^- > 0$.

(c) In period $t-1$ scenario $j \in J_{s,t-1}$ is selected and in period t scenario $j' \notin J_{st}$ is selected. Consequently, $\sum_{j' \in J_{st}} y_{tj'}$ does not exist. Hence, $x_{st}^- = 1$ (school closure).

(d) In period $t-1$ scenario $j \notin J_{s,t-1}$ is selected and in period t scenario $j' \in J_{st}$ is selected. Consequently, $\sum_{j \in J_{s,t-1}} y_{t-1,j}$ does not exist. Hence, $x_{st}^+ = 1$ (school opening).

5.3 Outlook

As a critique on this approach one has to mention that it does not allow explicitly discrete or continuous capacity variation. Moreover, there is no possibility to reallocate students before graduation. In other words, a school is closed, if the last student cohort has graduated. This is due to the fact that we are considering freshers (new enrollment) only. Future research should focus on model formulations which overcome these issues. Likewise an analysis of different model formulations for student allocation and its effects on scenario selection would be of interest. A new formulation for the scenario selection could be the maximization of student utility subject to a budget constraint.

6 Application to the City of Dresden

In this chapter we take Dresden as an example for an urban region with a temporary large decline in student numbers and apply the DSNPP to this urban area. First we discuss the data (Section 6.1) followed by the discussion of the results (Section 6.2) and some final remarks (Section 6.3).

6.1 Data for the DSNPP

Figure 6.1 shows the development of 5th grade college students from 1995 to 2020 in each district[1]. The maximum capacity of each college can be seen in Figure 6.2. We depend on information about the fix location costs per year for each college. Unfortunately, there exist no exact information about these costs. Hence, we use the replacement value related to the remaining life time of a location to approximate fix location costs. The investment costs per year for each school (modernization for instance) are added to the replacement value yielding periodised fix location costs for each college (see Figure 6.2).

[1]We choose districts (64) for visualization purposes, since the block level (6400) used for computation is too detailed.

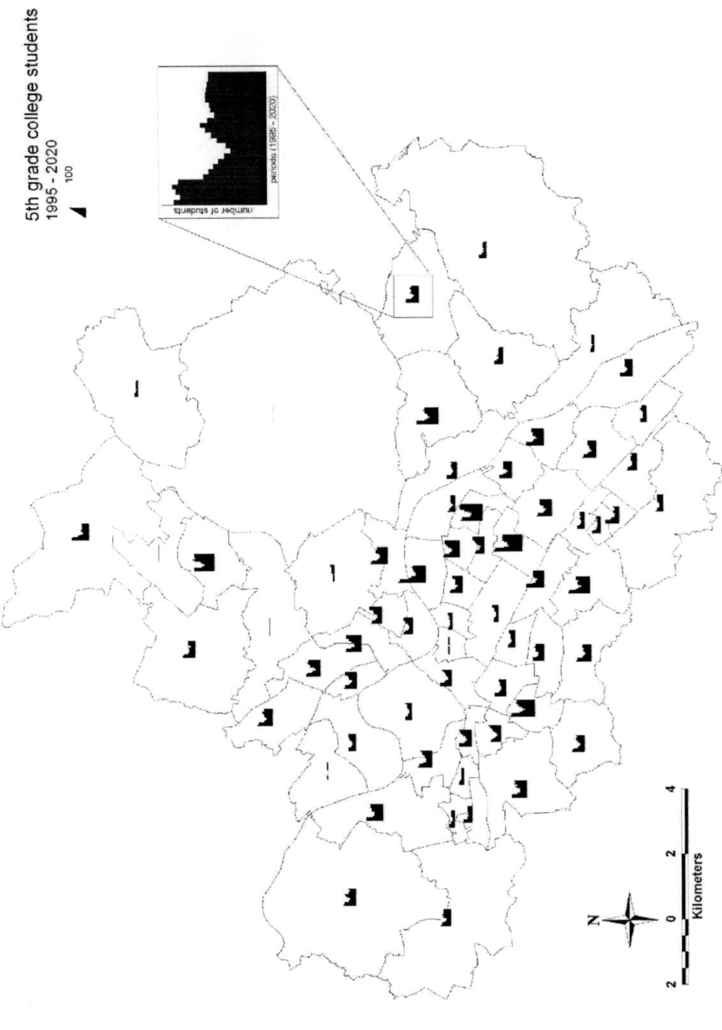

Figure 6.1: Number of College Students (5th grade) in Dresden from 1995 to 2020 (District Level)

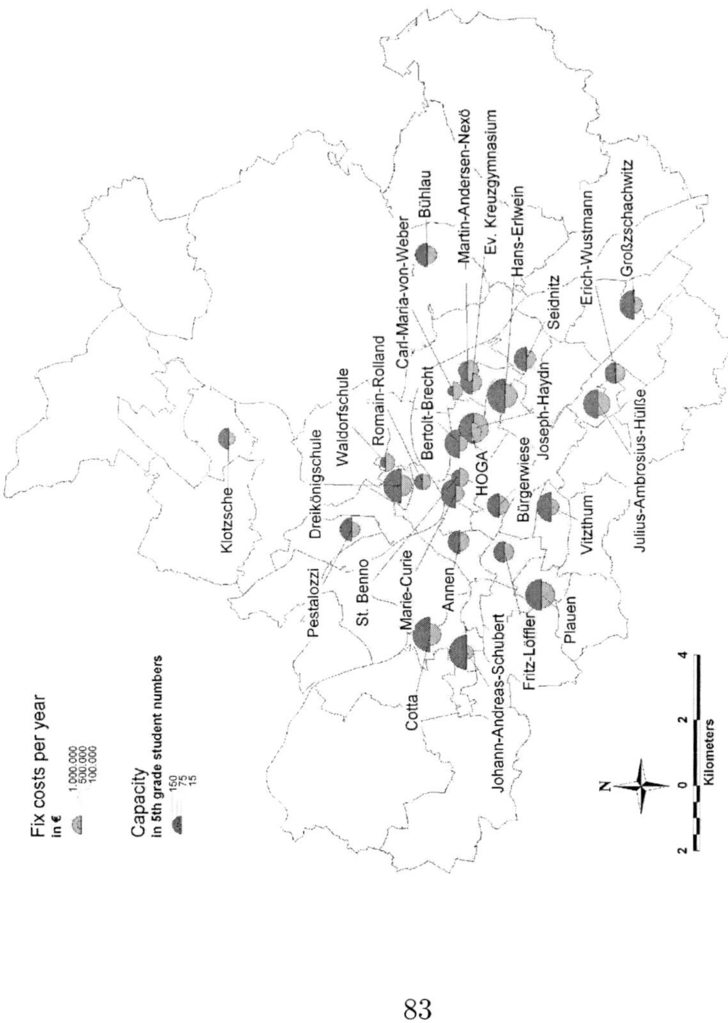

Figure 6.2: Capacity Provided for Students Located in Dresden (5th grade) and Periodised Fix Location Costs for Each College

The share of students to enroll on a given college is determined by MMNL2 specified in Section 3.2. The variable costs are set to 850 € per student per year for all colleges (Günther 2005). This includes operating costs for water, electricity etc. Moreover, for each student exceeding the maximum capacity we assume costs of 1105 € reflecting investment in capacity. The penalty costs per student below the minimum utilization are 1105 €. The weighting parameter $\alpha = 1$, while $\lambda = 1/|I|$. Since our objective is to consider the costs for the local authorities there are no fix and variable costs for private schools. Additionally, we do not consider individual transport costs and external diseconomies of transport, only compensation payments of 165 € per student and year by authorities for students of grades 5-10 using public transport services for distances larger than 3.5 km (City Council of Dresden (=Landeshauptstadt Dresden) 1997). The travel-to-school mode choice probabilities are determined by the MNL specified in Section 4.1. The discounting factor is set to 1.4 per cent which is the average inflation rate from 1995 to 2004. We assume the costs for the opening of a school are 1 million €[1] while closing a school causes handling costs of 100 000 €. Since we want to model the decision-making within the responsibility of the local education authority we have fixed private schools to be open over the whole planning horizon. Furthermore, we have fixed magnet schools (Romain-Rolland and Martin-Andersen-Nexö) since these are rather attractive as well as Plauen, Cotta, Dreikönigschule, Carl-Maria-von-Weber [2] and Klotzsche which are locally very important. So we assume that it would be challenging within a political context to close these colleges.
The fixing of schools yields a reduced number of possible scenarios and

[1]We assume that no new facilities will be built up, but other facilities like middle school buildings will be modified to be used as colleges. For a new building the costs per square meter are 500-700 € (see for example http://www.kleusberg.de/index/idart/670.html) and a college with 3 classes per grade has round about 3000 square meters (see for example http://www.bolle-modulbau.de/projekte/07-1.html).

[2]The responsible authority is the state Saxony and not the local education authority.

reduces computational effort. For example: We have 26 colleges and 25 periods. Fixing none of the colleges yields 2^{26} scenarios. When computing the student allocation (Section 5.2.1) we would have to consider $2^{26} \times 25$ possible combinations. Of course, some of these combinations are not feasible due to capacity reasons. We have fixed eleven colleges to be open over the whole planning horizon. This yields a maximum of 819 200 combinations of scenarios and periods.

6.2 Results

The student allocation has been computed with GAMS/Cplex 22.2 on a dual-core AMD Opteron 880 with 2.4 Ghz and 8 GB memory.[1]In order to simulate the school choice probabilities we set the number of draws to 250 to speed up computation. The computation time using two parallel threads is 7478 minutes. Based on the results, we consider 4 different alternatives for the DSNPP for scenario selection:

1. No variables are fixed.
2. y_{tj} is fixed to equal 1 for $t = 1995$ and the scenario $j \in J$ which represents the historical scenario in 1995.
3. The same as 2, but additionally $x^{+}_{st} = 0$ for $t < 2008$ for the colleges Bühlau, Seidnitz and Bürgerwiese. The local education authority decided in 2006 to open Bühlau in 2008 and the other two colleges later on. These locations are not available before 2008.
4. All y_{tj} are fixed in the way, that all decisions on school closures and openings of the local education authority are reproduced. Here, no model has to be solved. All costs can be computed directly.

[1]Data and source code is available on request.

Computation times with GAMS/Cplex 22.2 for the three alternatives of the DSNPP are - as expected - very different (see Table 6.1). The results of alternative 2 and 3[1] are exactly the same, so we just consider alternatives 1, 2 and 4. We use these solutions of the scenario selection (Section 5.2.2) to rerun the student allocation (Section 5.2.1) for the determined combinations of periods t and scenarios j (y_{tj}=1) with 1000 draws for school choice probability simulation to obtain enrollment and thus variable and transport costs as exact as possible. An overview of the results is given in Table 6.1. The first alternative with all variables free shows the best result followed by alternative 2 with a historical founded start scenario. The deviance between the political solution (alternative 4) and the optimal solution (alternative 1 and 2) yields the *political costs.* This is for the whole planning horizon between 1.731 million and 1.787 million € per year. We will consider the time span from 1995 - 2010 for a separate consideration, since the future decisions of the local education authority are not foreseeable. In this case the political costs are still from 1.678 million to 1.895 million € per year. Note that alternative 2 outperforms alternative 1 for the time span 1995-2010. This is due to the fix location costs (see below).

It is of great interest to find which schools are open or closed at certain points in time for a given alternative. Remember to interpret "closed" in the way that no 5th grade students are allowed to enroll at the considered school. Figures 6.3-6.5 show the resulting decisions on college openings and closings for all three alternatives. All schools which are not fixed are affected by opening or closure or even both (for different alternatives) except Vitzthum College, which is open over the whole planning horizon for all three alternatives. The decision of the local education authority (alternative 4) to close colleges seems to be related to the low number of students in the surrounding districts (see Figure 6.1): The closing of Johann-Andreas-Schubert, Erich-Wustmann and Großzschachwitz as well as the openings of Bühlau and Seidnitz are dispositional in this strat-

[1]Computation time: 163 minutes.

Alternative	1	2	4
1995-2020			
total costs in mio. €	433.88	435.28	478.56
relative deviation from alternative 1		.003	.10
relative deviation from alternative 2			.10
absolute deviation from alternative 1 (million €/year)		.055	1.787
absolute deviation from alternative 2 (million €/year)			1.731
1995-2010			
total costs in mio. €	297.53	294.27	322.70
relative deviation from alternative 1		-.01	.08
relative deviation from alternative 2			.10
absolute deviation from alternative 1 (million €/year)		-.217	1.678
absolute deviation from alternative 2 (million €/year)			1.895
number of closed colleges	9	8	6
CPU time in minutes	795	178	0

Table 6.1: Comparison of Alternatives Concerning Total Costs for Different Time-Frames, Time of Computation and The Number of Closed Colleges

egy. However, the optimal solution based on a historical start scenario (alternative 2) shows that it is inadvisable to close a lot of colleges located at the sparsely populated outskirts. It seems to be the better solution to thin out the spatial cluster of colleges located near the city center. This finding is confirmed by the results of alternative 1. However, at the same time it is suggesting to close certain colleges like Joseph-Haydn and Julius-Ambrosius-Hülße due to their high fix costs. Nevertheless, if these colleges are kept open they show relatively good utilization (see Appendix B). The optimal solutions (alternative 1 and 2) show better utilization than the political solution (alternative 4). This result is related to the strategy, not to close facilities located at the outskirts, but to close locations within a spatial cluster of schools (city center).

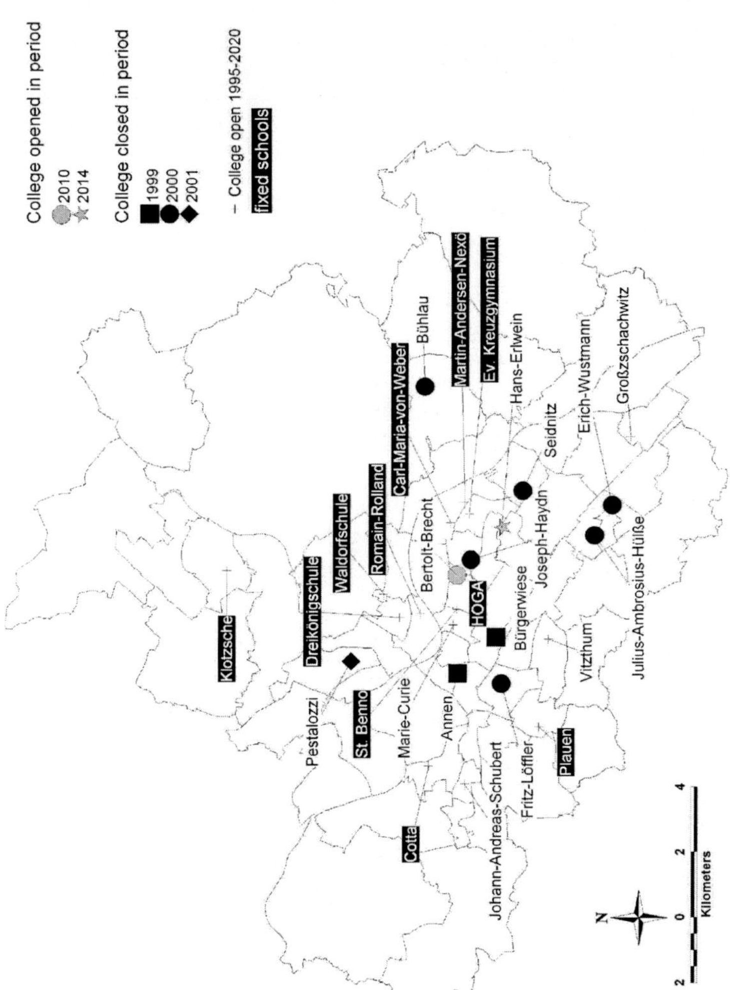

Figure 6.3: School Network of Alternative 1

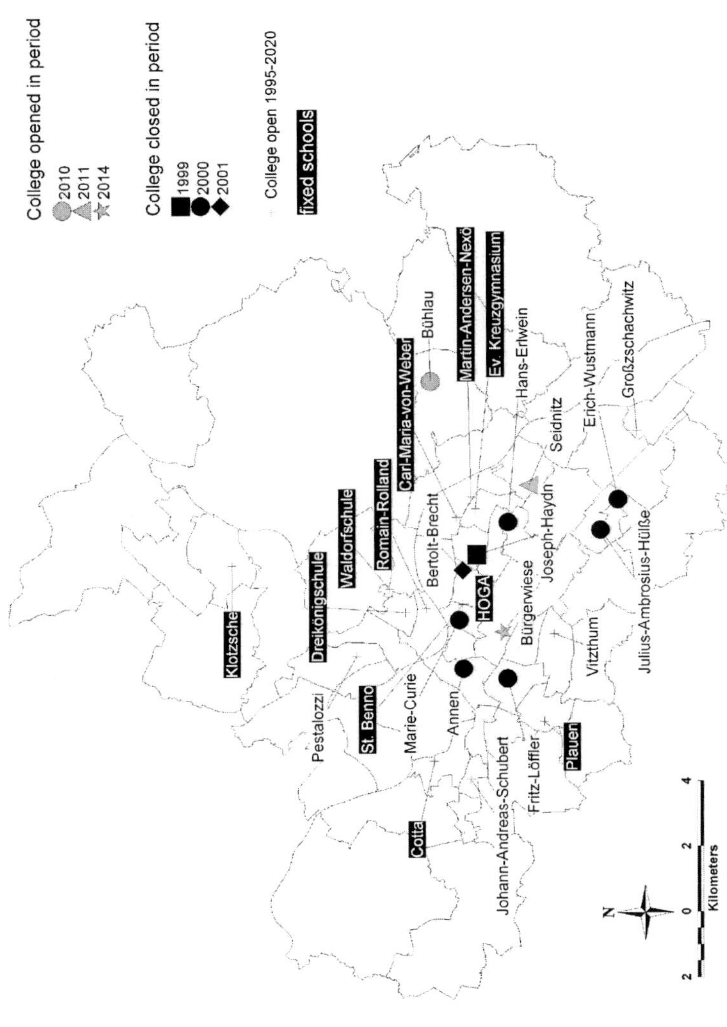

Figure 6.4: School Network of Alternative 2

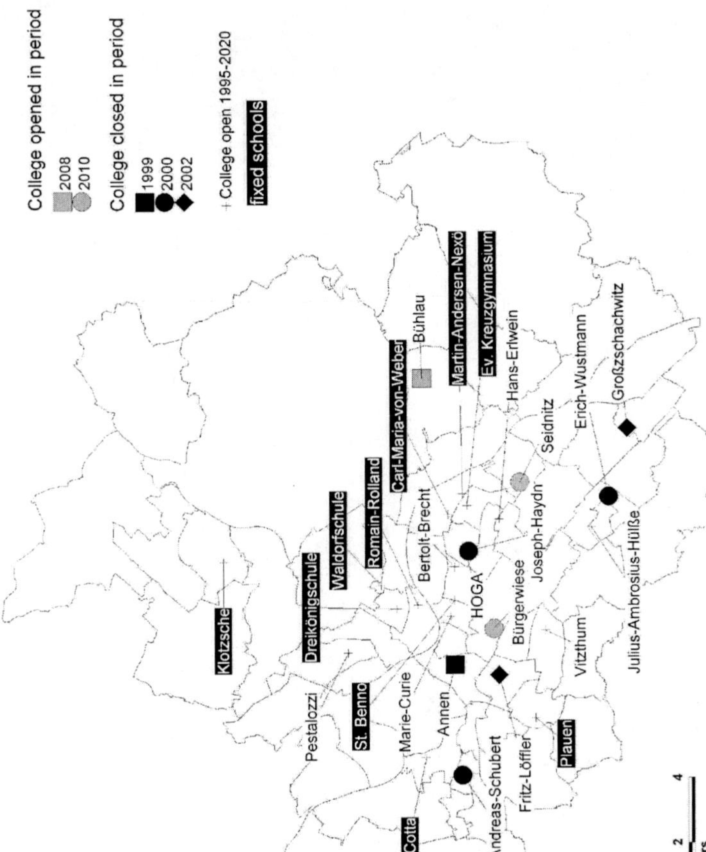

Figure 6.5: School Network of Alternative 4

The different costs of the alternatives can be seen in Figure 6.6. The total costs of the "political" alternative (alternative 4) differ remarkable from the costs of the optimal alternatives due to two reasons: First, the optimal alternatives feature more schools closed (see Table 6.1). And second, different decisions are made on the opening of colleges. This becomes more obvious if we just take the fix costs into account. From 2007 on the fix costs of alternative 4 are considerable above the fix costs of the optimal solutions. The fix costs of alternative 1 are far higher than that of alternative 2 for the time span 1995-1999. From 2000 to 2010 they are almost the same. Thus, alternative 2 outperforms alternative 1 for the planning horizon 1995-2010 (see Table 6.1).

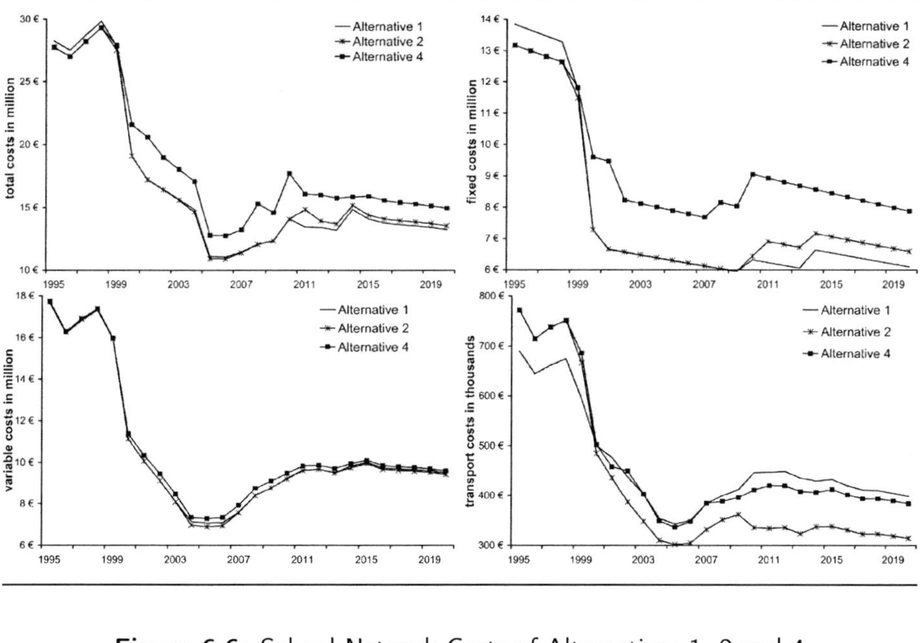

Figure 6.6: School Network Costs of Alternatives 1, 2 and 4

The education authority decided to open Bühlau in 2008 and Bürgerwiese and Seidnitz in 2010. In contrast, the optimal solutions incorporate the opening of Bertold-Brecht in 2010 as well as Hans-Erlwein in 2014 (alternative 1) and Bühlau in 2010, Seidnitz in 2011 and Bürgerwiese in 2014 (alternative 2). So, the peaks later than 2005 in total costs are related to the fix costs of newly opened colleges as well as to the opening costs. Compared to the fix costs and the total costs, the variable costs differ only a slightly between the alternatives. This is because, these costs depend on the number of attending students. Nevertheless, there is an observable discrepancy between alternative 1 and alternative 4 for the years 2000 - 2015. Besides the cost for attending students, there are penalty costs included in variable costs for violating minimum and maximum capacity constraints (Section 5.2.1). The penalty costs of alternative 4 due to overcapacity preponderates the penalty costs due to undercapacity of alternative 1.

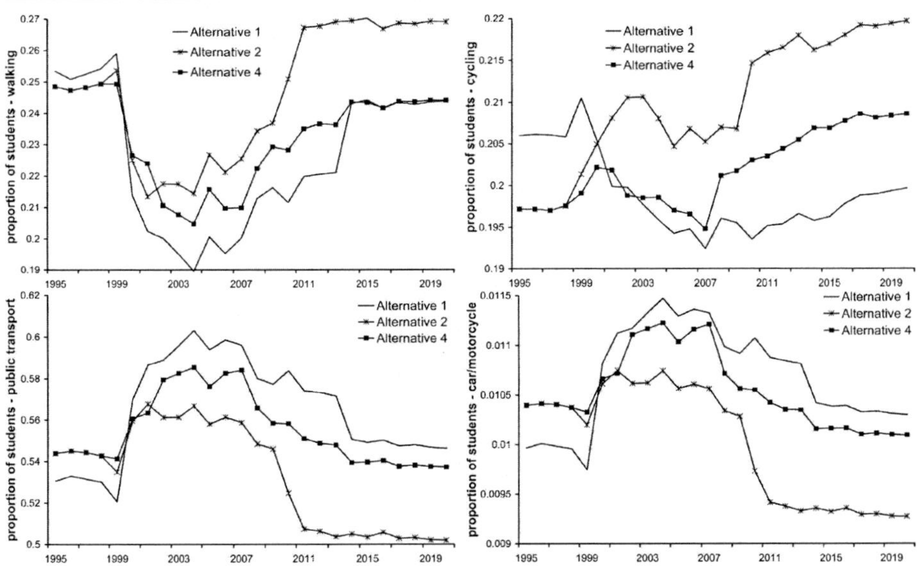

Figure 6.7: Modal Split of Alternatives 1, 2 and 4

Until 2000 the transport costs for alternative 1 are remarkable below the alternatives with a fixed (start) scenario. For instance the schools of alternative 1 are more easy to access by cycling and walking than the schools of alternatives 2 and 4 (see Figure 6.7). This school network is optimal adjusted to the spatial-temporal patterns of student location until the year 2000. After the years of school closing (1999-2001) we see a different picture. Now, alternative 2 outperforms alternative 1 and 4 - particular when the student numbers are increasing again. Less colleges are closed for alternative 4 and therefore the average distance to school is lower than this is the case for alternative 1. So alternative 4 yields less transport costs. But the spatial-temporal pattern of the school network for alternative 2 suits better than of alternative 4 although it contains less schools: It is remarkable that the opening of schools leads to reduced transport costs for alternative 2 but the opening of the same colleges in different periods yields increased transport costs for alternative 4.

6.3 Outlook

There is a need for faster methods for the student allocation. Moreover, if it is possible to speed up this process, one can use simulation to cope with uncertainty in student number forecasts. The development of an easy to use decision support system by incorporating this approach in a GIS may lead to widespread application by education and planning authorities.

7 Conclusions

With declining student numbers school authorities are forced to close school locations accompanied by changing choice sets for students. So it is worthwhile to know about the impacts on school choice behavior. As already stated in the literature we have found that distance has the strongest influence on school choice. But our analysis yields significant influence of profiles, authority responsible, centrality, and school size on school choice as well. Unique profiles and school size cause a remarkable surplus in utility. Hence, a specialization on a certain subject (like languages) will increase choice probability. While competition rises with declining student numbers, the consequence of these results is to focus on specific profiles in order to gain a given number of students. However, from the point of view of educational policy it is questionable whether such a development is desirable. By reason of changing choice sets over time it is important to model realistic substitution patterns. Therefore, we have used the flexible MMNL, which enables to analyse heterogeneity within population as well. Our analysis revealed significant taste variation for private schools, distance and centrality.

The analysis of the journey-to-school illustrates that weather or season respectively and car availability, have a strong influence on students' travel-to-school mode choice. It is revealed that distance is the most important factor for discrimination between commuting modes linked with high costs (public transport and car/motorcycle) and commuting modes linked with low costs (walking and cycling). However, we have seen exemplary even if we consider diseconomies of transport that the transport

costs do not outweigh the fix location costs. Note, these findings are only valuable for an urban area. The closure of a school location in rural areas will have much more dramatic effects on travel-times and modal-shift as well as other socio-economic consequences.
Most approaches for the DSNPP do not consider free school choice and geographical small scale real world applications. Therefore, we have presented a two-step, large scale optimization approach. The first step allocates the students to available school locations according to capacity and to maximize the students' utility. The student allocation has to be done for every feasible combination of period and scenario (scenario = set of available schools). This yields the scenario costs per period. In step two a scenario is selected for each period of the planning horizon according to minimize the scenario costs over all periods. We are able to determine the optimal time-space pattern for a school network (optimal period to open or close a facility location). Using the given data the results show that with an improvement of at least 8 to 10 per cent an amount of 1.5 to 2 million € per year could be saved. We interpret this gap between the optimal solution found and the political solution as political costs. In other words, the community is willing to pay these costs in order to get a desirable school network. Concerning the spatial consequences, the overall optimal solution guarantees a school network with minimum costs as well as best accessibility. We can say recapitulatory that our approach is an appropriate tool for decision support but it does not substitute political debate on the school network to choose.

A Appendix

Utilization of All Colleges Over Time for Each Alternative

College	Alternative 1	Alternative 2	Alternative 4
Annen	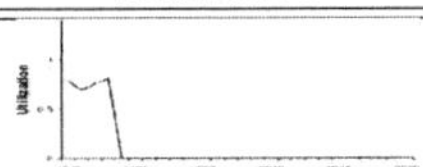	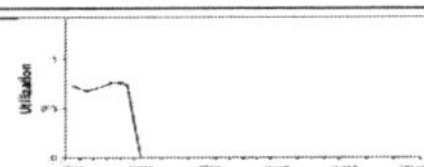	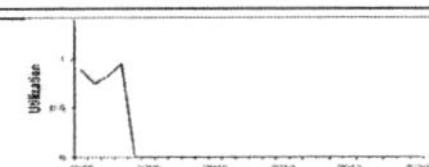
Bertold-Brecht	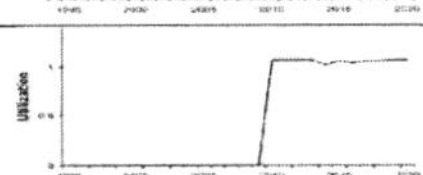	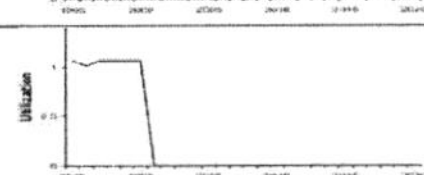	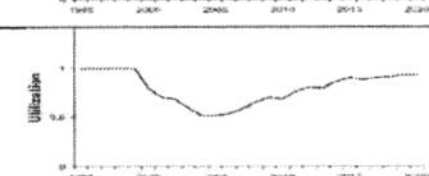
St.Benno			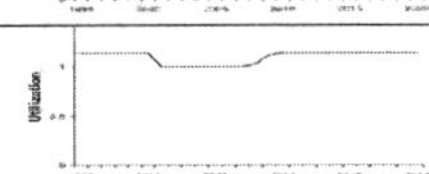
Bühlau	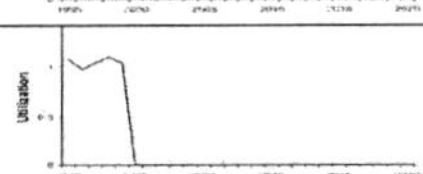	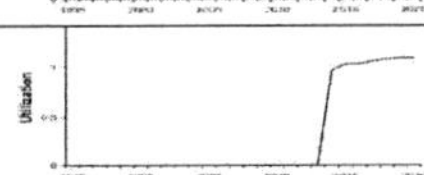	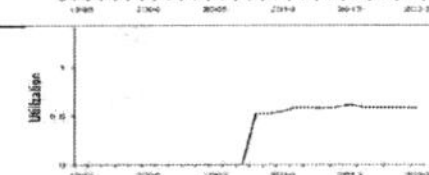
Carl-Maria-von-Weber	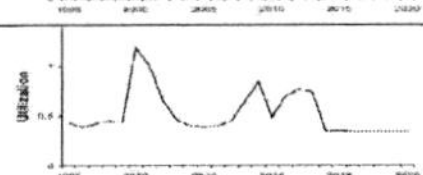	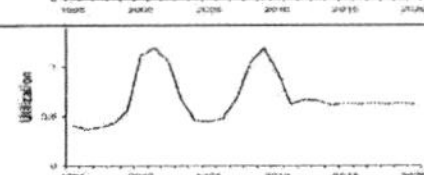	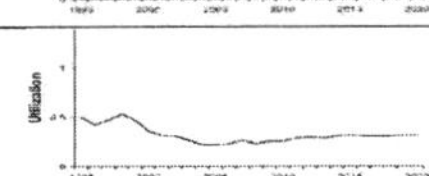
Cotta	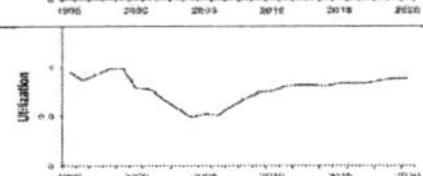	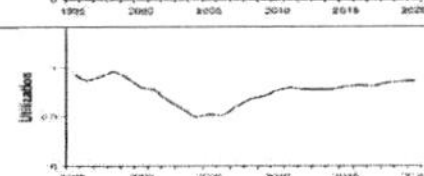	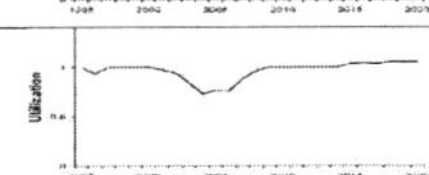
Dreikönigschule	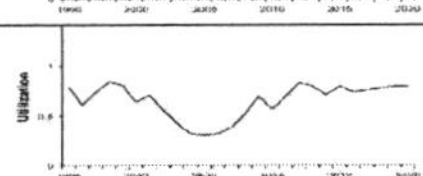	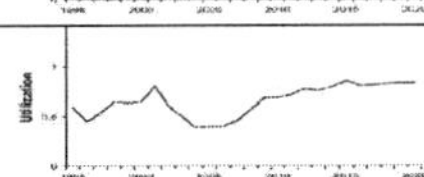	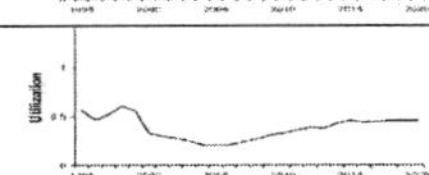

A Appendix

College	Alternative 1	Alternative 2	Alternative 4
Fritz-Löffler	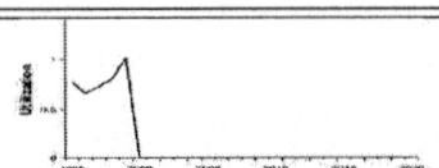	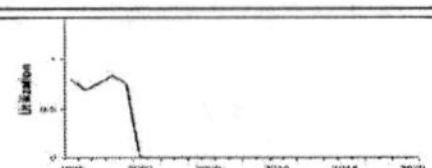	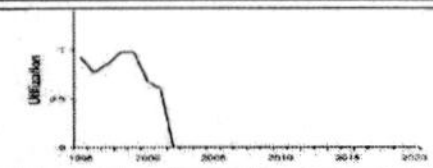
Großzschachwitz	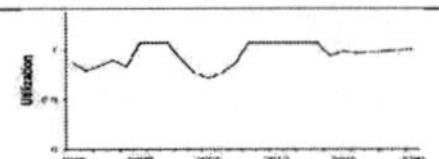	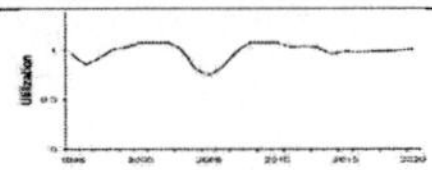	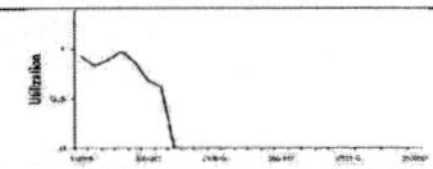
Hans-Erlwein	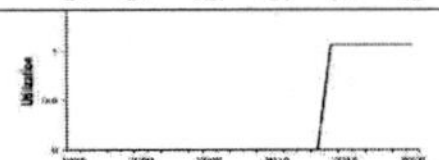	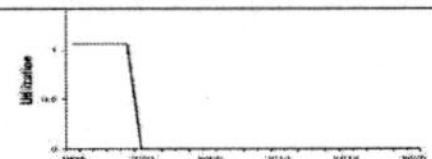	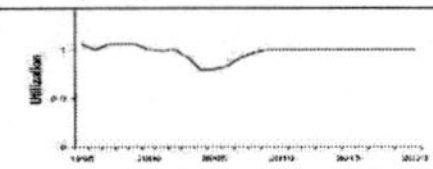
HOGA	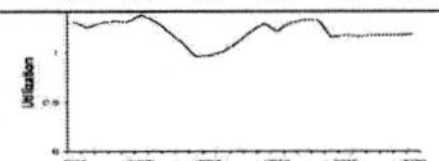	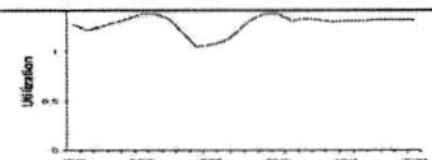	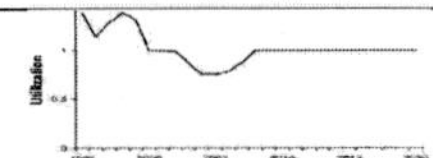
Julius-Ambrosius-Hülße	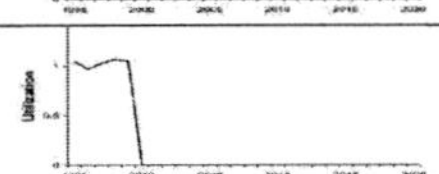	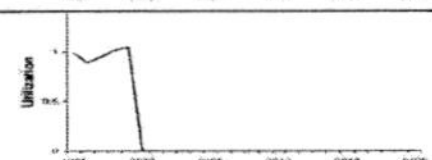	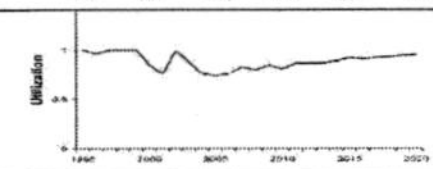
Johann-Andreas-Schubert	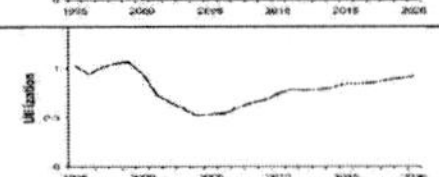	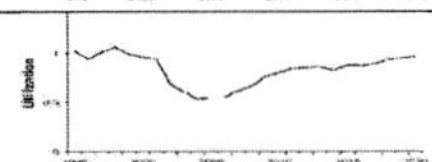	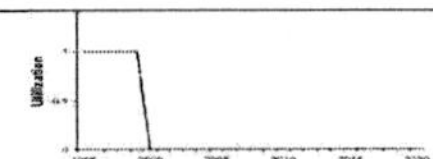
Joseph-Haydn	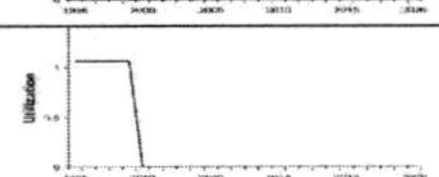	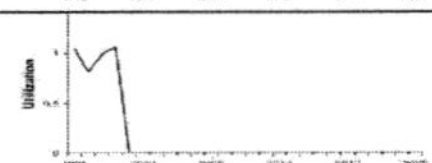	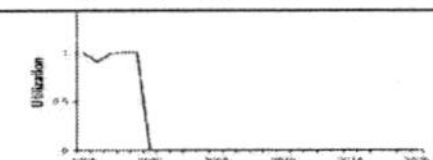
Klotzsche	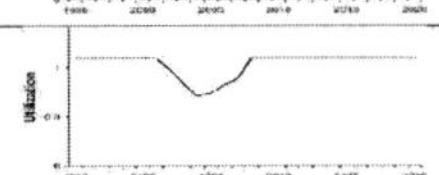	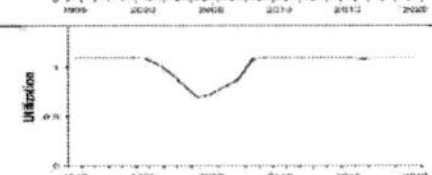	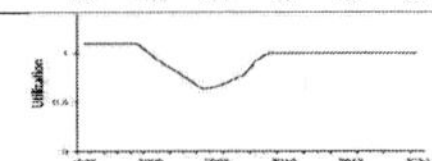
Ev. Kreutzgymnasium	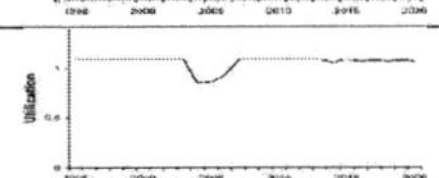	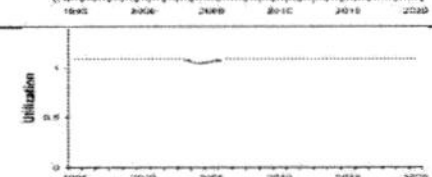	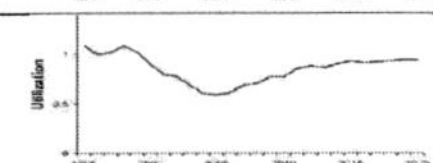
Martin-Andersen-Nexö	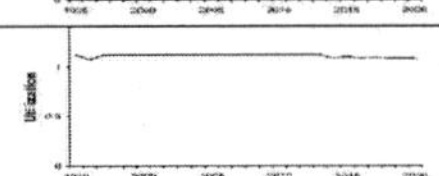		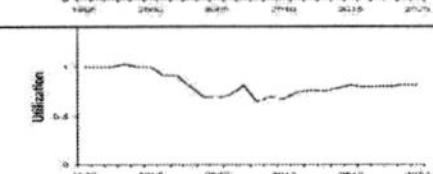
Marie-Curie	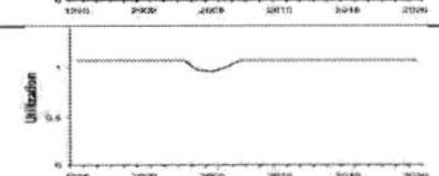	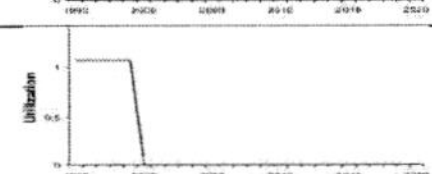	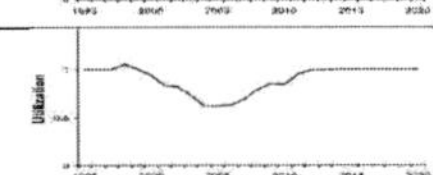

College	Alternative 1	Alternative 2	Alternative 4
Pestalozzi	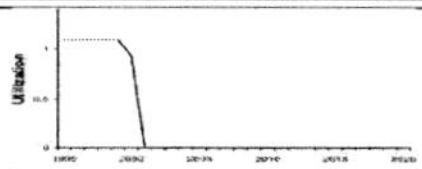	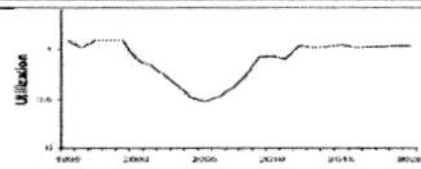	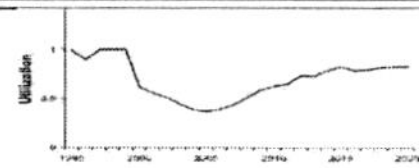
Plauen	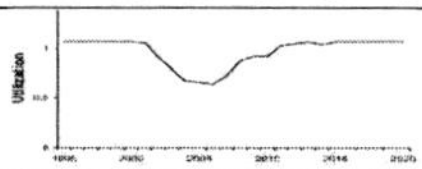	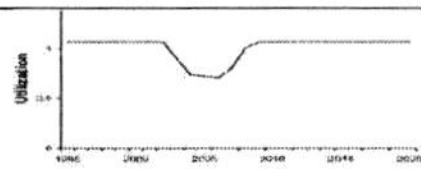	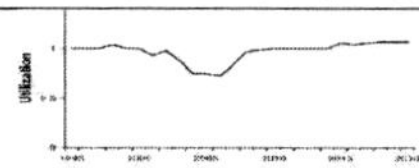
Romain-Rolland	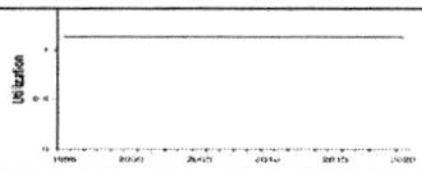	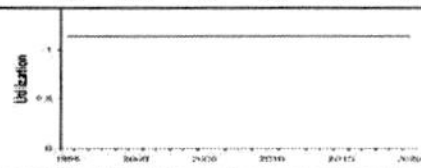	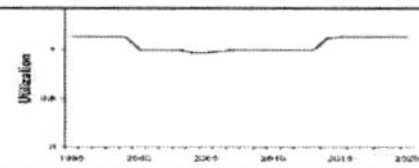
Seidnitz	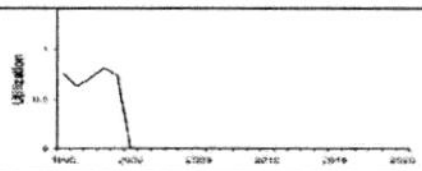	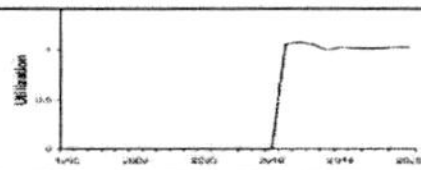	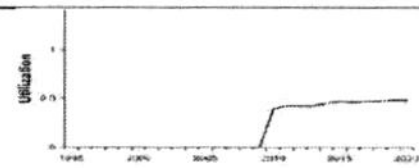
Vitzthum	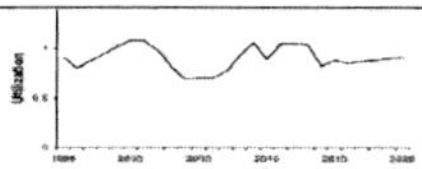	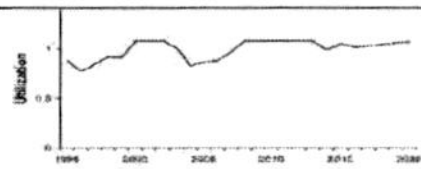	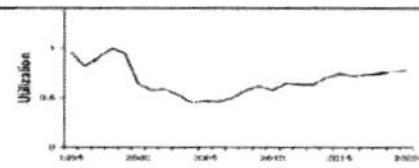
Waldorf	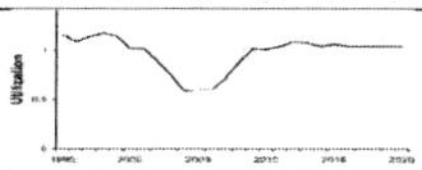	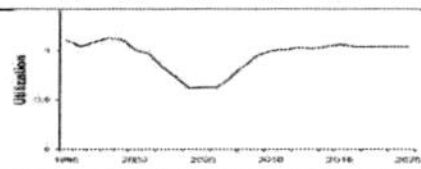	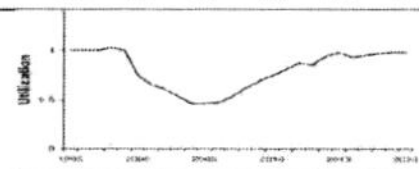
Bürgerwiese	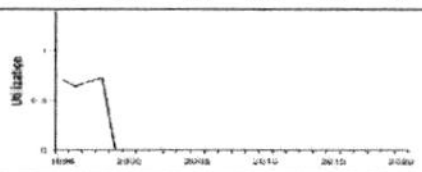	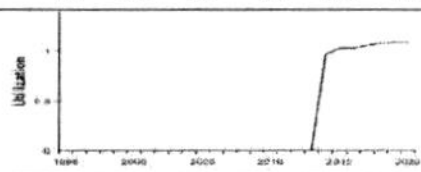	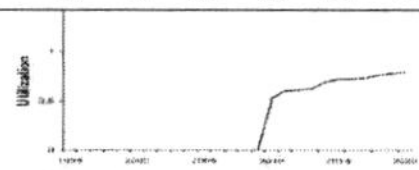
Erich-Wustmann	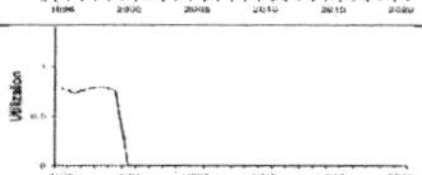	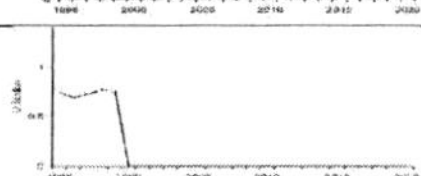	

References

Alderman, H., Orazem, P., Paterno, E., 2001. School quality, school cost and the public/private school choice of low income households in Pakistan. The Journal of Human Resources 36 (2), 304–326.

Antunes, A., Gomes, C., Ribeiro, A., 2004. An accessibility-maximization approach to public facility planning under distance-sensitive demands. Studies in Regional and Urban Planning 10, 109–117.

Antunes, A., Peeters, D., 2000. A dynamic optimization model for school network planning. Socio-Economic Planning Sciences 34, 101–120.

Antunes, A., Peeters, D., 2001. On solving complex multi-period location models using simulated annealing. European Journal of Operational Research 130, 190–201.

Axhausen, K., Haupt, T., Fell, B., Heidl., U., 2001. Searching for the rail bonus - results from a panel SP/RP study. European Journal of Transport and Infrastructure Research 1 (4), 353–369.

Bach, L., 1981. The problem of aggregation and distance for analyses of accessibility and access opportunity in location-allocation models. Environment and Planning A 13, 955–978.

Backhaus, K., Erichson, B., Plinke, W., Weiber, R., 2003. Multivariate Analysemethoden, 10th Edition. Springer, Berlin.

Bahrenberg, G., 1981. Providing an adequate social infrastructure in rural areas: an application of a maximal supply dispersion model to elementary school planning in Rotenburg/Wümme (FRG). Environment and Planning A 13, 1515–1527.

Barthels, K., 2007a. Sind Privatschulen besser? Die Zeit 43, http://www.zeit.de/2007/43/C–Privatschul–Info?page=all. 10. November 2007.

Barthels, K., 2007b. Wir leben Toleranz vor. Die Zeit 43, http://www.zeit.de/2007/43/C–Interview–Ev–Schule?page=all. 10. November 2007.

Bates, J., 2000. History of demand modelling. Vol. 1 of Handbooks in Transport (Handbook of Transport Modelling). Pergamon, Ch. 2, pp. 11–34.

Baum, H., Esser, K., Höhnscheid, K., 1998. Volkswirtschaftliche Kosten und Nutzen des Verkehrs. In: Forschungsarbeiten aus dem Strassen- und Verkehrswesen, Nr. 108.

Becker, A., Ekholm, K., Jäcle, R., Muendler, M., 2005. Location choice and employment decisions: A comparison of german and swedish multinationals. Review of World Economics 141 (4), 693–731.

Belford, P., Ratcliff, D., 1972. A network-flow model for racially balancing schools. Operations Research 20, 619–628.

Ben-Akiva, M., Lerman, S., 1985. Discrete choice analysis, theory and applications to travel demand. MIT Press, Cambridge, MA.

Bhat, C. R., 1997. A nested logit model with covariance heterogeneity. Transportation Research B 31, 11–21.

Bhat, C. R., 2000. Flexible model structures for discrete choice analysis. Vol. 1 of Handbooks in Transport (Handbook of Transport Modelling). Pergamon, Ch. 5, pp. 71–90.

Bhat, C. R., Gossen, R., 2004. A mixed multinomial logit model analysis of weekend recreational episodetype choice. Transportation Research B 38 (9), 767–787.

Bickel, P., Friedrich, R., 1995. Was kostet uns die Mobilität? Externe Kosten des Verkehrs. Springer, Heidelberg.

Bierlaire, M., 2003. BIOGEME: a free package for the estimation of discrete choice models. In: 3rd Swiss Transport Research Conference.

Bierlaire, M., 2005. An introduction to BIOGEME (Version 1.4).

Black, C., Collins, A., Snell, M., 2004. Encouraging walking: The case of journey-to-school trips in compact urban areas. Urban Studies 38 (7), 1121–1141.

Borgers, A., Oppewal, H., Ponjé, M., Timmermans, H., 1999. Assessing the impact of school marketing: conjoint choice experiments incorporating availablility and substitution effects. Environment and Panning A 31, 1949–1964.

Borges, O. M., Antunes, A., 2006. School network planning in Cap Vert, Maximizing accessibility vs minimizing investment. In: Joint International Conference on Computing and Decision Making in Civil and Building Engineering. Montreal, Canada, pp. 945–954.

Bovet, J., 1982. Simple heuristics for the school assignment problem. Journal of the Operational Research Society 33, 695–703.

Brinkmann, P., Leslie, L., 1987. Student price response in higher education. Journal of Higher Education 58 (2), 181–204.

Brownstone, D., Train, K. E., 1999. Forecasting new product penetration with flexible substitution patterns. Journal of Econometrics 89, 109–129.

Bruno, J. E., Andersen, P. W., 1982. Analytical methods for planning educational facilities in an era of declining enrollments. Socio-Economic Planning Sciences 16 (3), 121–131.

Burgess, S., Briggs, A., 2006. School assignment, school choice and social mobility. Working Paper 06/157, Center for Market and Public Organization.

Caro, F., Shirabe, T., Guignard, M., Weintraub, A., 2004. School redistricting: embedding GIS tools with integer programming. Journal of the Operational Research Society 55, 836–849.

Church, R. L., Murray, A. T., 1993. Modeling school utilization and consolidation. Journal of Urban Planning and Development 119 (1), 23–38.

Church, R. L., Schoepfle, O. B., 1993. The choice alternative to school assignment. Environment and Planning B 20, 447–457.

City Council of Dresden (=Landeshauptstadt Dresden), 1997. Satzung der Landeshauptstadt Dresden über die Erstattung der notwendigen Schülerbeförderungskosten, Satzung Schülerbeförderungskosten-Erstattung. Vol. 37 of Dresdner Amtsblatt.

City Council of Dresden (=Landeshauptstadt Dresden), 2003. Dresden in Zahlen. Tech. rep., Dresden.

Clarke, S., Surkis, J., 1968. An operations research approach to racial desegregation of school systems. Socio-Economic Planning Sciences 1, 259–272.

Conn, A., Gould, N., Toint, P., 2000. Trust region methods. In: MPS-SIAM Series on Optimization. SIAM.

Contreras, D., 2002. Vouchers, school choice and the access to higher education. In: Economic Growth Center Discussion Papers. Yale University.

de Boer, E., 2005. The dynamics of school location and school transportation: Illustrated with the dutch town of Zwijndrecht. TR News (237), 11–16.

Diamond, J. T., Wright, J. R., 1987. Multiobjective analysis of public school consilidation. Journal of Urban Planning and Development 113 (1), 1–18.

Dow, J. K., Endersby, J. W., 2004. Multinomial probit and multinomial logit, a comparison of choice models for voting research. Electoral Studies 23, 107–122.

Drezner, Z., Hamacher, H. W. (Eds.), 2001. Facility Location, Applications and Theory. Berlin.

Ewing, R., Forinash, C., Schroeer, W., 2005. Neighborhood schools and sidewalk connections: What are the impacts on travel mode choice and vehicle emission? Transportation Research News (237), 4–10.

Ewing, R., Schroeer, W., Greene, W., 2004. School location and student travel, analysis of factors affecting mode choice. Journal of the Transportation Research Board (TRB) (1895), 55–63.

Federal Environment Agency Germany (=Bundesumweltamt), 2007. Nachhaltige Entwicklung. http://www.umweltbundesamt.de/verkehr/verkehrstraeg/fussfahrad /texte/foerdnmiv.htm, 7.September 2007.

Federal Ministry of Transport, Building and Urban Affairs (=Bundesministerium für Verkehr, Bau und Stadtentwicklung (BMVBS)), 2002. Mobilität in Deutschland. http://www.mid2002.de/, 8.10.2006.

FGSV (Ed.), 2002. Wirtschaftlichkeitsuntersuchungen an Straßen: Stand und Entwicklung der EWS. Forschungsgesellschaft für Strassen- und Verkehrswesen (FGSV). Köln.

Figlio, D., Stone, J., 1997. School choice and student performance, are private schools really better? Vol. 1141-97 of Discussion Paper. Institute for Research on Poverty.

Fotheringham, A., Densham, P., Curtis, A., 1995. The zone definition problem in location-allocation modelling. Geographical Analysis 27 (1), 60 – 77.

Franklin, A., Koenigsberg, E., 1973. Computed school assignments in large districts. Operations Research 21, 413–426.

Gabriel, S., Rosenthal, S., 1989. Household location and race: Estimates of a multinomial logit model. The Review of Economics and Statistics 71 (2), 240–249.

German Aerospace Center (=Deutsches Zentrum für Luft- und Raumfahrt (DLR)), 2007. Verkehrslage mit floating-car-daten (fcd). http://www.taxifcd.de/Lage/HAMBURG/Barometer.html, 7.September 2007.

Gimona, A., Geddes, A., Elston, D., 2000. Localised areal disaggregation for linking agricultural census data to remotely sensed land cover data. Vol. 7 of Innovations in GIS, Computational Issues. Francis and Taylor, Philadelphia.

Glazerman, S., 1997. Determinants and consequences of parental school choice, unpublished working paper, Harris School of Public Policy, University of Chicago.

Goldring, E., Bauch, P., 1995. Parental involvement and school responsiveness: Facilitating the home-school connection in schools of choice. Educational Evaluation and Policy Analysis 17, 1–22.

Goodchild, M., 1979. The aggregation problem in location-allocation. Geographical Analysis 11 (3), 240–255.

Gottlieb, P. D., Joseph, G., 2006. College-to-work migration of technology graduates and holders of doctorates within the United States. Journal of Regional Science 46 (4), 627–659.

Greaves, S., 2004. GIS and the collection of travel survey data. Vol. 5 of Handbooks in Transport (Handbook of Transport Geography and Spatial Systems). Pergamon, Ch. 21, pp. 375–390.

Greene, W., 2003. Econometric Analysis, 5th Edition. Pearson Education, New Jersey.

Greenleaf, N. E., Harrison, T. P., 1987. A mathematical programming approach to elementary school facility decisions. Socio-Eonomic Planning Sciences 21, 395–401.

Günther, T., 2005. Gutachten zur Untersuchung der Sach- und Personalkosten öffentlicher Schulen und Schulen in freier Trägerschaft im Freistaat Sachsen. Tech. rep., Technische Universität Dresden.

Hanley, P. F., 2007. Transportation cost changes with statewide school district consolidation. Socio-Economic Planning Sciences 41, 163–179.

Hansen, E., 1987. Industrial location choice in Sao Paulo, Brazil. A nested logit model. Regional Science and Urban Economics 17, 89–108.

Hastings, J., Kane, T., Staiger, D., 2005. Parental preferences and school competition, evidence from a public school choice program. Tech. rep., National Bureau of Economic Reserach.

Haynes, K., Good, D., Dignan, T., 1988. Discrete spatial choice and the axiom of independence from irrelevant alternatives. Socio-Economic Planning Science 22 (6), 241–251.

Heckman, L., Taylor, H., 1969. School rezoning to achieve racial balance: A linear programming approach. Socio-Economic Planning Science 3, 127–133.

Heitkamp, S., 2005. Oft Schulalltag: Täglich drei Stunden im Bus. Dresdner Neueste Nachrichten 7./8. May 2005, 4.

Henig, M., Gerchak, Y., 1986. Dynamic capacity planning of public schools in changing urban communities. Socio-Economic Planning Sciences 20 (5), 319–324.

Hensher, D. A., Greene, W. H., 2003. The mixed logit model: The state of practice. Transportation 30 (2), 133–176.

Holloway, C. A., Wehrung, D. A., Zeitlin, M. P., Nelson, R. T., 1975. An interactive procedure for the school boundary problem with declining enrollment. Operations Research 23 (2), 191–206.

Hoxby, C. (Ed.), 2003. The economics of school choice. University of Chicago Press.

Hunt, J., Teply, S., 1993. A nested logit model of parking location choice. Transportation Research - B 27 (4), 253–265.

Infras/IWW, 2004. External Costs of Transport - Update Study, on behalf of International Union of Railways (UIC). Infras/IWW, Zürich/Karlsruhe.

Jennergen, L., Obel, B., 1980. A study in the use of linear programming for school planning in Odense. Journal of the Operational Research Society 31, 791–799.

Kane, J., Spizman, L., 1994. Race, financial aid awards and college attendance, parents and geography matter. American Journal of Economics and Sociology 53 (1), 85–95.

Klameth, S., 2005. Drei Stunden Schulweg. Sächsische Zeitung Online 10. June 2005.

Kleitz, B., Weiher, G., Tedin, K., Matland, R., 1998. Choices, charter schools and household preferences. Social Science Quarterly 81 (3), 846–854.

Koenigsberg, E., 1968. Mathematical analysis applied to school attendance areas. Socio-Economic Planning Sciences 1, 465–475.

Koppelmann, F., Sethi, V., 2000. Closed-form discrete choice models. Vol. 1 of Handbooks in Transport (Handbook of Transport Modelling). Pergamon, Ch. 13, pp. 211–226.

Koroglu, D., 1992. Possible locations for new universities in Turkey. Socio-Economic Planning Sciences 26 (1), 27–42.

Lankford, H., Wyckoff, J., 2000. The effect of school choice and residential location on the racial segregation of students, Working Paper.

Lemberg, D. S., Church, R. L., 2000. The school boundary stability problem over time. Socio-Economic Planning Sciences 34, 159–176.

Liggett, R., 1973. The application of an implicit enumeration algorithm to the school desegregation problem. Management Science 20, 159–168.

Lloyd, P. E., Dicken, P., 1990. Location in Space, Theoretical Perspectives in Economic Geography. Addison-Wesley.

Longley, P., Goodchild, M., Maguire, D., Rhind, D., 2001. Geographic Information Systems and Science. Wiley, Chichester.

Mahr-George, H., 1999. Determinanten der Schulwahl beim Übergang in die Sekundarstufe 1. Vol. 28 of Forschung Soziologie. Leske+Budrich, Opladen.

Maxfield, D., 1972. Spatial planning of school districts. Annals of the Association of American Geographers 62, 582–590.

McDaniel, R., 1975. Case study of the transportation algorithm for school districting under federal integration guidelines. Socio-Economic Planning Science 9, 271–272.

McDonald, M., 2005. Children's travel, patterns and influences. Ph.D. thesis, University of California, Berkeley.

McFadden, D., 1973. Conditional logit analysis of quantitative choice behaviour. In: Zaremmbka, P. (Ed.), Frontier of econometrics. Academic Press, New York.

McKeown, P., Workman, B., 1976. A study in using linear programming to design non-urban attendance areas. Socio-Economic Planning Science 10, 223–226.

McMillan, T., 2003. Walking and urban form: Modelling and testing parental decisions about children's travel. Ph.D. thesis, University of California, Irvine.

Molinero, M., 1988. Schools in Southampton: A quantitative approach to school location, closure and staffing. Journal of the Operational Research Society 39 (4), 339–350.

Montgomery, M., 2002. A nested logit model of the choice of a graduate business school. Economics of Education Review 21 (5), 471–480.

Müller, S., 2006. Analyse des Verkehrsmittel- und Schulwahlverhaltens Dresdner Gymnasiasten. In: Aurada, K., Rödel, R. (Eds.), Kooperation und Integration. Vol. 39 of Greifwalder Geographische Arbeiten. pp. 145–155.

Müller, S., Haase, K., 2007. Simulating school choice probabilities using a mixed multinomial logit model, working Paper.

Müller, S., Haase, K., Kless, S., 2007a. A dynamic school network planning approach with free school choice, working Paper.

Müller, S., Tscharaktschiew, S., Haase, K., 2007b. Travel-to-school mode choice modelling and patterns of school choice in urban areas. Journal of Transport Geography (accepted for publication).

Nguyen, A., Taylor, J., 2003. Post-high school choices: New evidence from a multinomial logit model. Population Economics 16 (2), 278–306.

November, S., Cromley, R., Cromley, E., 1996. Multi-objective analysis of school district regionalization alternatives in Connecticut. The Professional Geographer 48 (1), 1–14.

Oosterhaven, J., 2005. Spatial interpolation and disaggregation of multipliers. Geographical Analysis 37 (1), 69–84.

Peter, J., 2004. Das große Schulsterben. Die Welt 26. August 2004, 3.

Peter, J., 2005. Koalitionsstreit in Sachsen. Die Welt 20. May 2005, 4.

Peterson, P., Hassel, B. (Eds.), 1998. Learning from school choice. Brookings Institution, Washington, D.C.

Pizzolato, N. D., 1994. A heuristic for large-size p-median locations with application to school location. Annals of Operations Research 50, 473–485.

Planco Consulting GmbH (Ed.), 1993. Gesamtwirtschaftliche Bewertung von Verkehrswegeinvestitionen: Bewertungsverfahren für den Bundesverkehrswegeplan 1992. Essen.

Richter, K., 2005. Schwärzeste Prognosen übertroffen . Dresdner Neueste Nachrichten 31. May 2005, 11.

Rouwendal, J., Meijer, E., 2001. Preferences for housing, jobs, and commuting: A mixed logit analysis. Journal of Regional Science 41 (3), 475–486.

Schlottmann, K., 2005. Trend zur Mini-Schule. Sächsische Zeitung 10. November 2005, 12.

Schneider, T., 2004. Der Einfluss des Einkommens der Eltern auf die Schulwahl. Vol. 446 of Discussion Papers. German Institute for Economic Research (=Deutsches Institut für Wirtschaftsforschung (DIW)), Berlin.

Schoepfle, B. O., Church, R. L., 1989. A fast, network-based, hybrid heuristic for the assignment of students to schools. Journal of the Operational Research Society 40, 1029–1040.

Schoepfle, B. O., Church, R. L., 1991. A new network representation of a classic school districting problem. Socio-Economic Planning Sciences 25 (3), 189–197.

Smith, M. D., 2005. State dependence and heterogeneity in fishing location choice. Journal of Environmental Economics and Management 50, 319–340.

Speiser, I., 1993. Determinanten der Schulwahl, Privatschulen - öffentliche Schulen. Vol. 11 of Europäische Hochschulschriften 513, Pädagogik. Lang, Frankfurt a.M.

Spiekermann, K., Wegener, M., 2000. Freedom from the tyranny of zones: towards new GIS-based spatial models. In: Fotheringham, A., Wegener, M. (Eds.), Spatial models and GIS - New potential and new models. Taylor & Francis, London, pp. 45–62.

Stimson, D., Thompson, R., 1974. Linear programming, busing and educational administration. Socio-Economic Planning Science 8, 195–206.

Sutcliffe, C., Board, J., Cheshire, P., 1984. Goal programming and allocating children to secondary schools in Reading. Journal of the Operational Research Society 35 (8), 719–739.

Taylor, R. G., Vasu, M. L., Causby, J. F., 1999. Integrated planning for school and community: The case of Johnston County, North Carolina. Interfaces 29 (1), 67–89.

Teixeira, J., Antunes, A., Peeters, D., 2007. An optimization-based study on the redeployment of a secondary school network. Environment and Planning B 34, 296–315.

Train, K., E., 2003. Discrete choice methods with simulation. Cambridge University Press.

Trifon, R., Livnat, A., 1973. The spatial allocation of schools over time in cities. Regional and Urban Economics 2 (4), 387–400.

Tsai, J., Miller, M., 2005. Integrated planning for school and community. Transportation Research Record 1922.

Van der Horst, D., 2002. The benefits of more spatial detail in regional bioenergy models and the problems of disaggregating agricultural census data. Options Méditerranéennes A 48, 131–138.

Weber, A., 1909. Über den Standort der Industrien. Tübingen.

Woodside, A., Gunay, B., Seymour, J., 2002. Travel to school: a study of the effect of socio-economic differences in secondary schools on the choice of 'travel to school' modes. In: Proceedings European Transport Conference (PTRC). Cambridge.

Yeates, M., 1963. Hinterland delimination: a distance minimizing approach. The Professional Geographer 15, 7–10.

Stadt- und Raumplanung

Andreas Hofer; Klaus Semsroth;
Bohdan Tscherkes

Urbane Metamorphosen für die Krim

Die Krim ist ein Ort magischer Anziehungskraft zwischen Orient und Okzident. Für die Menschen im kalten Norden der Ukraine und Russlands liegt hier der Traum von Süden, Sonne und Strand. Die faszinierende Kulturlandschaft mit ihrem idealen Klima bildet seit mehr als hundert Jahren die Kulisse für Glück und Erholung: erst für die russische Aristokratie, dann für die Werktätigen des realen Sozialismus und heute für den freien Markt. Doch heute finden sich hier auch die postsowjetischen Städte mit ihrer heruntergekommenen Infrastruktur, verlassene Sanatorien und eine desorientierte Bevölkerung, die nach Arbeit sucht. Dieses Buch berichtet von der Architektur und den Städten auf der Krim, von deren Geschichte und Perspektiven, sowie vom Alltag und den Träumen der dort lebenden Menschen. Eine Publikation des Instituts für Städtebau Landschafts- architektur und Entwerfen der Technischen Universität Wien

Bd. 1, 2005, 184 S., 14,90 €, br., ISBN 3-8258-9202-6

Katja Veil

Sicherheit im Wohnquartier und Stadtplanung

Herausforderungen und Perspektiven am Beispiel ethnischer Minderheiten in Großbritannien

Gerade im Bereich der Sicherheit wird die Stadtplanung häufig auf den Aspekt der städtebaulichen Gestaltung fokussiert. Das Ziel dieses Buches ist, aus der Perspektive der städtebaulichen Kriminalprävention herauszutreten und die gesamte Raumdimension von Sicherheit in den Fokus der Stadtplanung zu stellen. In einer Fallstudie wird der räumliche Zusammenhang zwischen geringer relativer Sicherheit und ethnischer Zugehörigkeit in Großbritannien als eine Herausforderung für planerisches Handeln dargestellt. In Anschluss daran werden planerische Steuerungsmöglichkeiten diskutiert.

Bd. 4, 2008, 256 S., 24,90 €, br., ISBN 978-3-8258-1716-9

LIT Verlag Berlin – Hamburg – London – Münster – Wien – Zürich
Fresnostr. 2 48159 Münster
Tel.: 0251 / 620 32 22 – Fax: 0251 / 922 60 99
e-Mail: vertrieb@lit-verlag.de – http://www.lit-verlag.de